Springer Hydrogeology

The Springer Hydrogeology series serves as a comprehensive platform for advancing knowledge and fostering collaboration in the diverse and multidisciplinary field of hydrogeology. This book series seeks to cater to a wide audience, including researchers, educators, practitioners, policymakers, and students who are passionate about understanding and solving groundwater-related challenges. It aims to provide high-quality, peer-reviewed monographs, edited volumes, textbooks, and conference proceedings to disseminate cutting-edge research, practical insights, and innovative approaches in hydrogeology and related disciplines.

The series covers the entire spectrum of hydrogeological topics, including but not limited to: groundwater science and management, isotope hydrogeology, hydrogeological modeling, groundwater resources and systems, hydrogeochemistry, ecohydrology and groundwater-energy-food nexus, cryohydrogeology, urban and agricultural hydrogeology, emerging technologies in hydrogeology, climate change impacts on groundwater and groundwater-dependent ecosystems, and geotechnical and environmental aspects.

The Springer Hydrogeology series is aligned with the United Nations Sustainable Development Goals (SDGs), particularly SDG 6 (Clean Water and Sanitation) and SDG 13 (Climate Action). By publishing works that address sustainable groundwater use, pollution mitigation, climate resilience, and integrated groundwater resource management, the series supports efforts to achieve equitable and sustainable access to groundwater resources for current and future generations. Contributions that explore the intersection of hydrogeology with other SDGs—such as those on biodiversity, sustainable cities, and food security—are also strongly encouraged.

Peiyue Li • Jianhua Wu • Xiaodong He
Yuanhang Wang • Xiaofei Ren • Lingxi Li

Water Resources Research in Northern Qinling Mountains, China

Springer

Peiyue Li
School of Water and Environment
Chang'an University
Xi'an, China

Jianhua Wu
School of Water and Environment
Chang'an University
Xi'an, China

Xiaodong He
School of Water and Environment
Chang'an University
Xi'an, China

Yuanhang Wang
School of Water and Environment
Chang'an University
Xi'an, China

Xiaofei Ren
School of Water and Environment
Chang'an University
Xi'an, China

Lingxi Li
School of Water and Environment
Chang'an University
Xi'an, China

ISSN 2364-6454 ISSN 2364-6462 (electronic)
Springer Hydrogeology
ISBN 978-3-032-09394-3 ISBN 978-3-032-09395-0 (eBook)
https://doi.org/10.1007/978-3-032-09395-0

This Springer imprint is published by the registered company Springer Nature Switzerland AG
The registered company address is: Gewerbestrasse 11, 6330 Cham, Switzerland

Preface

The Qinling Mountains, stretching across Central China, are more than a geographic landmark, but they are also a hydrological keystone that sustains ecological and human systems on both sides of the Yangtze River and Yellow River. Often referred to as China's "central water tower," the Qinling Mountains play a pivotal role in regulating the nation's water security, biodiversity, and climate dynamics. Among its many functions, the northern flank of the Qinling Mountains serves as a vital recharge and conservation zone for the Guanzhong Plain, which houses densely populated cities like Xi'an. This unique transitional belt is not only rich in surface and groundwater resources, but is also under increasing stress from urban expansion, agricultural intensification, and climate variability.

Despite decades of scientific exploration, water resources research in the northern Qinling Mountains still faces critical challenges. Integrated understanding of surface–groundwater interactions, ecological thresholds under anthropogenic pressure, and the resilience of the water system under climate change remains limited. Complex terrain, data scarcity, and the lack of unified management approaches across administrative and disciplinary boundaries have hindered progress in securing the region's long-term water sustainability. This book, *Water Resources Research in Northern Qinling Mountains, China*, emerges from recent national scientific expeditions and research projects, aiming to bridge these knowledge gaps with field-based insights, modeling tools, and interdisciplinary analyses. It is hoped that this work will not only advance the science of hydroecology in the Qinling region but also inform water governance practices and environmental planning in Central China.

This book has eleven chapters. The first nine chapters lay a foundational understanding, from broad geographic and hydrological contexts to detailed studies of rivers, groundwater, ecological flow requirements, water quality, land use changes, and nitrate pollution risks. They create a coherent narrative that progresses from baseline characterization to applied assessment and management strategies. The second last chapter (Chap. 10) integrates these insights into a holistic framework for sustainable water resource management and ecological protection in the Qinling Mountains, while the last chapter (Chap. 11) summarizes the main conclusions in this book and proposes future research directions.

This opening chapter (Chap. 1) sets the stage by framing the Qinling Mountains as a crucial geographic and ecological boundary with far-reaching hydrological significance. It emphasizes the northern Qinling's role as the primary water source for the Guanzhong Plain, outlining the scientific challenges and knowledge gaps. By summarizing past research and recent fieldwork, the chapter defines key questions that guide the book.

Chapter 2 offers a comprehensive overview of the northern Qinling region's physical and human dimensions. It details the area's complex climate and topography, tectonic history, and rich biodiversity, as well as socio-economic factors like urbanization, agriculture, tourism, and mineral resources. Highlighting the growing imbalance between water supply and demand, the chapter stresses the region's vulnerability amid rapid economic development.

Focusing on the critical surface water systems, Chap. 3 examines the more than 250 rivers originating from the northern slopes, categorizing them based on geomorphology and recharge sources. Detailed case studies illustrate the hydrological functions, water quality, ecological roles, and infrastructure developments across key river basins. This chapter underscores the rivers' fundamental importance to urban water supply and ecological conservation, laying the groundwork for assessing ecological base flows and water quality issues discussed in the subsequent chapters.

Recognizing the threats posed by seasonal drying and reduced runoff, Chap. 4 investigates ecological base flow estimation methods to safeguard river health. Employing both hydraulic and hydrological approaches, it quantifies spatial variations in base flow and examines how flow interruptions relate to climatic, hydrogeological, and human factors. The chapter concludes by proposing targeted monitoring and management strategies, bridging hydrological science with practical conservation needs and informing policy discussions on ecological water requirements.

Water quality is a vital component of ecosystem and human health. Chapter 5 analyzes surface water chemistry through extensive field sampling and advanced statistical and isotopic methods. It identifies key drivers of water quality variability and evaluates suitability for drinking, irrigation, and industry. Highlighting challenges such as pollution and habitat degradation, the chapter offers scientific insights essential for environmental protection policies and sustainable water management in the northern Qinling region.

Complementing the surface water focus, Chap. 6 addresses the hydrochemical characteristics and quality status of groundwater, which is critical for local water supply and ecological balance. The chapter reveals dominant chemical types, pollution hotspots, especially nitrate contamination linked to agriculture, and overall water suitability for human use. The chapter provides a crucial understanding of groundwater's role and vulnerabilities, emphasizing the need for integrated surface and subsurface water management.

Chapter 7 explores the vital connectivity between surface water and groundwater systems in three key river basins, using isotopic, hydrochemical, and statistical techniques. Understanding these interactions is fundamental for managing watershed water resources, protecting riparian ecosystems, and preventing pollution spread. By linking hydrological and geochemical data, this chapter bridges the gap between surface and subsurface studies, highlighting the integrated nature of the regional water cycle.

Human activities are transforming the landscape, influencing water circulation and quality. Chapter 8 analyzes two decades of land use change in a selected area of the northern Qinling Mountains and projects future trends using remote sensing and modeling. It documents decreases in farmland, increases in urban areas and water bodies, and the resulting implications for ecological conservation and sustainable development. This land-water nexus perspective is critical for anticipating future water challenges and guiding regional planning.

Focusing on nitrate pollution, Chap. 9 investigates its spatial distribution, sources, and associated health risks within a critical subregion. Employing advanced statistical and isotopic models, it identifies dominant nitrate sources and quantifies non-carcinogenic health risks for local populations. The chapter emphasizes the influence of land use patterns, particularly agriculture and residential areas, on groundwater contamination, offering evidence-based recommendations for mitigating pollution and protecting public health.

Finally, Chap. 10 synthesizes these insights into actionable water resource protection and management strategies tailored to the complex socio-ecological realities of the northern Qinling Mountains. Chapter 11 summarizes the main findings of this book and discusses future perspectives.

The Qinling Mountains stand like a majestic and profound chronicle, with every fold inscribed by the codes of civilization spanning thousands of years. The waters flowing from Qinling have nurtured the splendid and enduring Chinese civilization. Qinling's water resources are intimately linked to China's present development, yet their benefits extend far into the future, bearing the responsibility of sustainable development for generations to come. These waters are the foundation of our very existence and the cornerstone of Qinling's own sustainable future. In the face of mounting environmental challenges and growing societal demands, we need now more than ever to deepen our understanding and scientifically manage the water resources of this vital land. Only by pooling wisdom and strength across all sectors, and by conducting systematic and comprehensive research on Qinling's water resources, can we safeguard this lifeline and realize harmony between ecology and economy. Let us join hands to protect Qinling's water resources, ensure regional water security, and promote green development, thereby writing a new chapter in the protection and utilization of Qinling's water heritage in this new era.

Xi'an, China

Peiyue Li
Jianhua Wu
Xiaodong He
Yuanhang Wang
Xiaofei Ren
Lingxi Li

Acknowledgments

The publication of this book, Water Resources Research in the Northern Qinling Mountains, China, has been made possible through the support and encouragement of many individuals and organizations. We would like to express our deepest gratitude to all those who contributed to its completion.

Our Qinling research began in 2019, with the support of the Shaanxi Provincial Committee of the Communist Youth League of China (CYLC). With their generous backing, we were awarded six Qinling Scientific Expedition Projects, which enabled us to carry out continuous research on water resources in the Qinling Mountains. We sincerely thank CYLC for providing us with the opportunity to contribute to the scientific exploration of this vital region.

This book is the result of collaborative effort and teamwork. In addition to the listed authors, many master's and Ph.D. students have participated in extensive field investigations. Most of them have graduated from Chang'an University, but their valuable contributions to the research will never be forgotten. We would especially like to acknowledge the efforts of Fengmei Su, Song He, Abel Nsabimana, Dan Wang, Fei Xu, Xiaofei Li, Jing Ning, Qixiao Zhang, Yujie Ji, Miaojun Wei, Zhenmin Su, Yanan Guo, Yanlin Li, Lei Zhang, Wenqu Li, Leining Liu, Rui Duan, Lei Wang, Wenyu Guo, Chunliu Yang, Qianqian Du, Ningning Yang, Yongqiang Yang, Yuhan Zhou, Fan Li, Jiahui Li, Yinfu Chen, Weichao Liu, Yueyue Meng, and Linjie Xie. Their dedication and hard work have been instrumental in bringing this project to fruition.

This work was also supported by several research grants, including those from the National Natural Science Foundation of China (42472316, 42072286, 41761144059, and 41502234), the National Key Research and Development Program of China (2023YFC3706901), the Fundamental Research Funds for the Central Universities (300102295201), the Natural Science Basic Research Program of Shaanxi Province (2025JC-YBQN-375 and 2024JC-YBQN-0303), the Qinchuangyuan "Scientist + Engineer" Team Development Program of the Shaanxi Provincial Department of Science and Technology (2022KXJ-005), the National Ten Thousand Talents Program (W03070125), and the Fok Ying Tong Education Foundation (161098). These grants provided the essential resources for our research and the production of this book.

We hope this book will serve as a valuable resource for researchers, policymakers, and practitioners working in the field of water resources. Furthermore, we aspire for it to help the global community better understand the water systems of the Qinling Mountains. It is our sincere wish that this work will inspire continued collaboration and innovation in addressing the challenges of water security amid climate change, land use transformations, and other human-induced pressures.
School of Water and Environment, Chang'an University

Xi'an, China

Peiyue Li
Jianhua Wu
Xiaodong He
Yuanhang Wang
Xiaofei Ren
Lingxi Li

Competing Interests The authors have no competing interests to declare that are relevant to the content of this manuscript.

Contents

About the Authors

Peiyue Li is a Full Professor of Hydrogeology and Environmental Sciences and the Vice Dean of School of Water and Environment at Chang'an University, China. He holds a Ph.D. in Groundwater Science and Engineering (2014) and possesses wide experience in groundwater quality assessment, hydrogeochemistry, and groundwater modeling. Peiyue Li has been selected as a Young Talent for the National Ten Thousand Talents Program and recognized as a Young Changjiang Scholar by the Ministry of Education. He has edited and co-edited 12 textbooks and academic monographs and has published over 220 articles in refereed journals. He currently serves as an Associate Editor for *Exposure and Health, Mine Water and the Environment, Archives of Environmental Contamination and Toxicology, Environmental Monitoring and Assessment, Human and Ecological Risk Assessment, Applied Water Science, Water Practice and Technology, and Discover Water.* He is also a Book Series Editor for *Springer Hydrogeology, Cave and Karst Systems of the World, Advances in Karst Science, and GeoHealth: The Impact of Geology on Environment and Human Health*. He has been awarded Highly Cited Researcher status by Clarivate, Elsevier, and other organizations.

Jianhua Wu is a Full Professor at Chang'an University in China, specializing in hydrology and water resources in arid areas, loess seepage and its disaster mechanisms, and groundwater environmental protection. Her research has been supported by numerous projects funded by the National Natural Science Foundation of China. Her research output includes over 80 academic papers published in international journals, with 16 of them recognized as highly cited and popular papers by Essential Science Indicators. In addition to her research contributions, Dr. Wu actively engages with the academic community as an Associate Editor for the international journal *Exposure and Health*.

Xiaodong He is an Assistant Professor at Chang'an University, China. He holds a Ph.D. in hydraulic engineering from Chang'an University and possesses extensive expertise in hydrogeochemistry, water-rock interaction, water ecology, and water security. He has published over 20 scientific papers in renowned international journals, focusing on groundwater quality and pollution. In recognition of his research excellence, he was selected as one of the Top 2% Scientists Worldwide by Stanford University in 2022.

Yuanhang Wang is a young researcher with extensive experience in hydrogeology and environmental science. His research interests encompass a wide range of topics, including groundwater quality assessment, groundwater hydrochemistry, moisture movement, and in situ tracer tests. Many of his papers have gained considerable recognition and have been widely cited by fellow researchers. He successfully chaired the Doctoral Foundation project of the Ministry of Education. Additionally, he has actively participated in numerous national and provincial research projects.

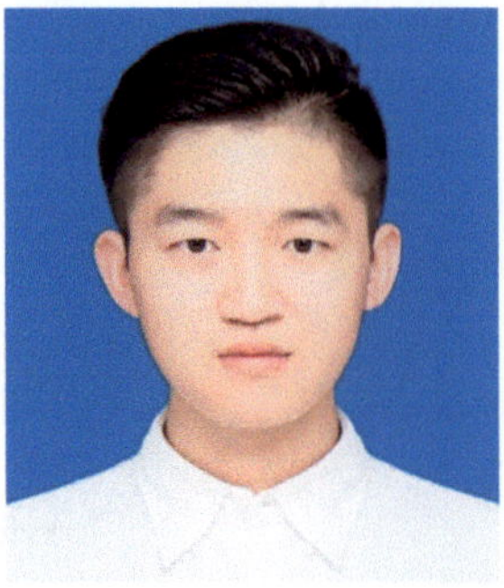

Xiaofei Ren is a young researcher who has made significant contributions to the fields of hydrogeology and environmental science. His work includes groundwater quality assessment, groundwater hydrochemistry, isotope hydrology, and ecological base flow. Dr. Ren's involvement in national and provincial research initiatives, particularly the National Natural Science Foundation Project, showcases his active engagement in advancing scientific knowledge.

Lingxi Li is a researcher affiliated with the School of Water and Environment at Chang'an University. Her research interests revolve around the assessment of groundwater and surface water environmental quality, as well as hydrogeochemistry. Lingxi Li's contributions to the field have been recognized by her receipt of provincial awards twice for her work on the Qinling Ecological and Environmental Protection Projects.

Chapter 1
Water Research in Qinling Mountains: An Introduction

Abstract This is an introductory chapter and it introduces the Qinling Mountains as a pivotal geographic, ecological, and hydrological boundary in central China, highlighting their critical role as the nation's central water tower. It outlines the unique significance of the northern Qinling region as a key source of water for the Guanzhong Plain. The chapter reviews past research on tectonics, biodiversity, hydroclimatology, and water resources, revealing both the richness of existing knowledge and persistent scientific gaps such as insufficient understanding of surface–groundwater interactions, climate impacts on water retention, and the ecological thresholds of anthropogenic stress. Drawing on results from recent field studies and national research projects, this chapter defines the objectives and structure of the book. Key questions to be addressed include hydrological dynamics, pollution risks, ecological sustainability, and integrated water resource strategies.

Keywords Water resources · Climate change · Environmental pollution · Ecological risk · Water management

1.1 Significance of the Qinling Mountains

The Qinling Mountains stand prominently in the center of China, and is one of the world's "Three Great Central Mountain Ranges" alongside the European Alps and the North American Rocky Mountains. The name "Qinling" originates from the ancient Qin Dynasty in China, with its earliest record in Shiji (Records of the Grand Historian).

As the geographical divide between northern and southern China, the Qinling Mountains govern the regions between the Yellow River to the north and the Yangtze River to the south (Fig. 1.1). They serve as a vital ecological shield for central and western China, playing a prominent role in the nation's economic framework. The Qinling Mountains are also an iconic central orogenic belt, climatic boundary, and ecological barrier. They are regarded as the ancestral pulse of the Chinese nation

P. Li et al., *Water Resources Research in Northern Qinling Mountains, China*, Springer Hydrogeology, https://doi.org/10.1007/978-3-032-09395-0_1

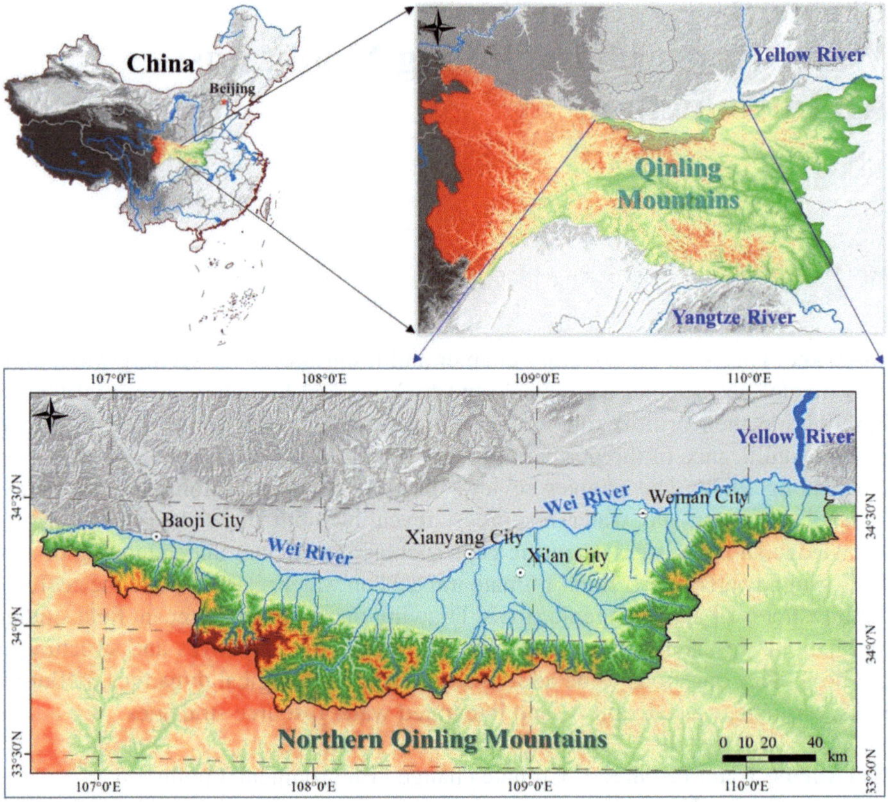

Fig. 1.1 Geographical location of the Qinling Mountains

and a significant symbol of Chinese culture, holding unparalleled importance for Chinese cultural heritage.

The Qinling Mountains function as China's central water tower, providing critical ecological services such as water conservation, climate regulation, and biodiversity maintenance. With abundant water reserves, they are a major water source for the Yellow River Basin, the Yangtze River Basin, and the central route of the South-to-North Water Diversion Project, influencing and regulating water security across northern and southern China (Liu et al. 2024; Zhang et al. 2025). In a broader sense, the Qinling Mountains stretch approximately 1600 km from east to west, from Baishishan in Lintan County, Gansu Province, to the Funiu Mountains in Henan Province. They span 200–300 km from north to south, covering an area of about 400,000 km^2. However, in a narrower sense, the Qinling Mountains refer to the section within Shaanxi Province, extending 400–500 km east-west and 100–150 km north-south, with an average elevation of 1500–2500 m and a total area of about 80,000 km^2. In this book, the northern Qinling Mountains refer to the transitional zone between the Qinling Mountains and the Wei River, stretching from the main ridge of the Qinling Mountains to the Wei River. This area spans a vast region across

Baoji, Xi'an, and Weinan cities in Shaanxi Province, China (Fig. 1.1). This area is a vital part of the Guanzhong Basin (also known as Guanzhong Plain).

Within Shaanxi Province, the Qinling Mountains hold approximately 22.2 billion m^3 of water resources, accounting for about 50% of the province's total water reserves. The northern Qinling Mountains contribute roughly 4 billion m^3 of water, representing about 51% of the surface water resources in the Guanzhong Basin. This area is a crucial water conservation zone for the Guanzhong Basin and a vital water source for the Guanzhong urban agglomeration. It is the lifeline of the Guanzhong water ecosystem, sustaining ecological water systems, water quantity, and economic development. Water is the foundation of ecology, and the protection of the Qinling Mountains' ecological environment is inseparable from water resources.

The northern Qinling Mountains region is bordered by the Qinling Mountains to the south and the Wei River to the north, situated in the heart of the Guanzhong Plain. This region is the most densely populated area with concentrated villages and towns and frequent production and construction activities. However, it faces ecological and environmental challenges such as weak water resource protection, insufficient ecological functions in some rivers, and an inadequate flood prevention system. Therefore, it is urgent and necessary to do comprehensive water research in this region.

1.2 Research on Qinling Mountains

The northern Qinling Mountains region exhibits co-evolutionary characteristics in terms of tectonics, geomorphology, environment, and ecosystems (Dong et al. 2022). Given the region's significance, numerous scholars have conducted research on its ecosystems and biodiversity, geological evolution and disaster prevention, water resources and climate change, cultural heritage, and regional development.

The Qinling Mountains are an iconic central orogenic belt, formed through long-term, multi-stage orogenic movements. They represent a composite continental orogenic belt and serve as the tectonic suture zone between the North China and South China blocks, forming the backbone of the Chinese mainland (Peng et al. 2023). This region preserves a complete record of granitic magmatism from subduction (250–235 Ma) to continental collision (235–210 Ma). Widespread Triassic magmatic activity has produced massive granitoid bodies with asymmetrical distribution and abundant mafic microgranular enclaves (MMEs). A significant proportion of mineral deposits in the region are closely associated with these Triassic granitoid intrusions, making the Qinling orogenic belt an ideal natural laboratory for studying the relationship between lithospheric structure and metallogenesis (He et al. 2025b). By comparing vertical fault slip rates and erosion rates along the northern Qinling piedmont fault zone, Shi et al. (2021) found that the northern Qinling topography has experienced continuous uplift and growth since the Late Pleistocene. The northeastward growth of the Tibetan Plateau has modulated modern erosion in the northern Qinling Mountains, shaping its present-day topography.

The Qinling Mountains are also a distinctive climatic boundary and ecological barrier. The region hosts over 3800 species of seed plants, 418 bird species, and 112 mammal species, accounting for 13%, 29%, and 22.4% of China's total species, respectively. Among these, 120 animal and 176 plant species are listed as national or provincial key protected species. The giant panda, golden snub-nosed monkey, takin, and crested ibis are collectively known as the "Four Treasures of the Qinling Mountains". The forests of the Qinling Mountains cover approximately 80.4% of the total area, with a forest coverage rate of 82% (data as of 2022). The region also harbors 30 mineral resources with reserves ranking among the top ten in China, including molybdenum, nickel, vanadium, barite, and limestone. Additionally, the Qinling Mountains are home to over 600 medicinal plants, such as *Salvia miltiorrhiza*, *Eucommia ulmoide*s, and *Gynostemma pentaphyllum*, earning it the title of China's "Natural Medicine Treasury" (Peng et al. 2023).

The ridgeline of the Qinling Mountains aligns with China's 800 mm isohyet, 2000-h sunshine isoline, and the January 0 °C isotherm, marking it a significant hydrological divide between northern and southern China. The mountains act as a barrier to air exchange between the north and south and are highly sensitive to climate change. Studying the response mechanisms of tree growth in the Qinling region to climatic factors is crucial for predicting trends in the region's forest ecosystems. Li et al. (2022) established a standardized tree-ring width chronology (STD) for *Pinus tabuliformis* (Chinese Red Pine) on the southern slopes of Nanwutai in the northern Qinling Mountains to investigate the correlation between radial growth and climatic factors. Meng et al. (2024) examined the spatiotemporal response of vegetation coverage to extreme climate events in the Qinling Mountains across multiple time scales. They found that the spatial patterns of annual and seasonal average NDVI (Normalized Difference Vegetation Index) were consistent between the northern and southern slopes, showing higher values in the central region and lower values at the edges. The maximum consecutive 5-day precipitation in spring and spring precipitation were the most significant positive controlling factors for vegetation. Fan et al. (2014) developed a mechanistic model using climatic variables to study the impact of landscape-scale climate change on the spatial distribution of giant panda habitats in the Qinling Mountains, and assessed changes in habitat distribution and range under future climate scenarios. The Qinling Mountains are a central hub for forest carbon sinks and a core oxygen-producing region in China. In 2015, the forest carbon sink in the Qinling region totaled approximately 6.78 GtC, accounting for 7.04% of the national total, while oxygen production reached 1063 billion tons, representing 8.66% of the national total. Gong et al. (2024) investigated the impact of climate change on carbon and water fluxes in the Qinling Mountain ecosystem. Their findings indicated that the Qinling Mountains consistently played a role in water conservation and carbon sequestration under changing climatic conditions.

The Qinling Mountains are China's central water tower, with their northern foothills providing a continuous supply of clean water to the Guanzhong region. Shen et al. (2025) mapped the spatial distribution and developmental characteristics of valleys in the northern Qinling foothills and explored their relationship with the Guanzhong water system using Google Earth remote sensing imagery, DEM data, and river network vector data, combined with multi-modal spatial recognition and data interpretation in ArcGIS and Global Mapper. Xu et al. (2024) proposed a typological framework for valley roads in the northern Qinling foothills based on the concept of ecological–economic–social collaborative development. He et al. (2025a) evaluated soil and water conservation in the northern Qinling foothills using the InVEST model and random forest analysis, and identified key influencing factors. They found that water retention capacity was significantly higher in mountainous areas than in the alluvial plains to the north of the Wei River. Key factors affecting soil and water conservation included annual precipitation, potential evapotranspiration, forest-to-agricultural land ratio, and soil texture. Precipitation, forest coverage, and topsoil sand content positively influenced water retention, whereas evapotranspiration, agricultural land, and topsoil clay content had negative effects. Yu et al. (2014) used stable oxygen isotopes to reveal distinct moisture sources between the northern and southern slopes of the Qinling Mountains. He et al. (2021) compared runoff variations and their causes between the northern and southern foothills of the Qinling Mountains. Zhang et al. (2022) predicted runoff trends in the Qinling Mountains and analyzed their impact on water demand in the Guanzhong region. Jia et al. (2020) used multiple isotopes and chemical tracers to identify nitrate sources in shallow groundwater to address groundwater pollution caused by intensive human activities (e.g., industrial, agricultural, and domestic pollution) in the northern Qinling Mountains.

Scholars have conducted studies on groundwater resource development and utilization, groundwater resource and vulnerability assessments, dynamic characteristics of groundwater levels and influencing factors, hydrogeochemical evolution, surface water–groundwater interactions, and precipitation response relationships in the northern Qinling regions (Wang et al. 2020, 2023; Yuan et al. 2024). However, key scientific gaps remain, including the lack of integrated models coupling surface water, groundwater, and valley morphology dynamics; unquantified climate change impacts on long-term water tower resilience; undefined ecological thresholds for anthropogenic pollution; fine-scale spatiotemporal variability in water retention, unclear groundwater recharge mechanisms under valley development, unassessed hydrological impacts of eco-engineered valley roads, and limited predictive frameworks using AI-driven multi-source data fusion. Additionally, nature-based solutions and cross-subbasin water management synergies between the northern and southern foothills warrant further exploration to balance ecological preservation and human needs. These unresolved problems highlight the need for interdisciplinary approaches to safeguard the Qinling's critical water resources.

1.3 Objectives and Scope of This Book

This book synthesizes the key findings from the Qinling Scientific Expedition Projects implemented during 2019–2022 and some provincial and national research projects over the past 5 years. Based on these projects, extensive fieldwork, sampling, laboratory analyses, and model calculations/simulations were conducted. These efforts enabled us to complete comprehensive investigations into river discharge and water quality, groundwater quantity and quality, land use patterns, and ecological conditions across the area. This book will be a valuable resource for water professionals and practitioners seeking new knowledge about water resources in this rarely reported area of China. It will also provide meaningful insights for decision-makers involved in formulating effective water policies.

1.3.1 Purpose of the Book

This book was written to meet the following purposes:

- **Systematize Hydrological Knowledge**: Compile fragmented research on the northern Qinling's water resources into a cohesive framework, integrating field data, remote sensing, and assessment results.
- **Address Critical Gaps**: Bridge unresolved scientific questions, such as surface-groundwater interactions, baseflow sustainability, and nitrate pollution under land-use change identified during the expedition.
- **Support Policy and Management**: Provide evidence-based strategies for water resource protection, aligning with China's ecological conservation goals for the Qinling Mountains.
- **Promote Interdisciplinary Collaboration**: Highlight the nexus of hydrology, ecology, and socioeconomics to guide future research and sustainable development in this ecologically vital region.

1.3.2 Key Questions Addressed in This Book

Water research in the north Qinling Mountains is critical for the sustainable development in this area. This book attempted to address the following key questions:

- **Hydrological Dynamics**: How do the northern Qinling's unique topography and climate shape river baseflow and groundwater recharge? What are the spatiotemporal patterns of surface water–groundwater exchange across major river basins?
- **Water Quality Challenges**: What are the dominant pollutants in surface water and groundwater, and how do their sources (e.g., agriculture, industry) vary

spatially? How does land-use change (e.g., urbanization, deforestation) exacerbate nitrate contamination and health risks?
- **Ecological Sustainability**: Can current ecological baseflow calculations ensure the health of riparian ecosystems amid climate change and human demand? What are the thresholds of anthropogenic pressure (e.g., pollution, water extraction) beyond which water tower functions degrade?
- **Management Solutions**: Which integrated strategies (e.g., eco-engineering, policy reforms) can balance water resource use with conservation in the Guanzhong region's context?

By answering these questions, the book aims to translate expedition insights into actionable science for preserving the Qinling's role as China's "central water tower".

1.3.3 Overview of the Book's Structure

This book is structured to provide a systematic and comprehensive examination of the region's hydrological systems, addressing both scientific and management challenges. The book begins with an Introduction (Chap. 1), outlining the significance of the Qinling Mountains, reviewing existing research, and defining the book's purpose, key questions, and structure. Chapter 2 offers a detailed overview of the study area, covering geography, climate, topography, geology, water resources, and socio-economic conditions.

The core of the book delves into the hydrology of the northern Qinling Mountains, starting with an analysis of the main rivers (Chap. 3), their characteristics, and regional distribution. Chapter 4 focuses on baseflow estimation, assessing ecological flow requirements and proposing strategies to sustain river health. Chapters 5 and 6 evaluate surface water and groundwater quality, respectively, using chemical and statistical methods to identify pollution sources and spatial patterns. Chapter 7 investigates surface water–groundwater interactions, employing multiple techniques to quantify recharge dynamics across different river basins.

The latter part of the book explores human impacts on water resources. Chapter 8 analyzes land use changes and their future projections, while Chap. 9 examines groundwater nitrate pollution in selected areas under shifting land use and assesses associated health risks. Finally, Chap. 10 synthesizes the findings into actionable water resource protection and management strategies, addressing challenges in water quantity, quality, and ecosystem sustainability.

By integrating hydrology, geochemistry, remote sensing, and policy analysis, this book provides a multidisciplinary perspective on the northern Qinling's water resources, serving as a vital reference for researchers, policymakers, and environmental managers.

References

Dong Y, Shi X, Sun S et al (2022) Co-evolution of the Cenozoic tectonics, geomorphology, environment and ecosystem in the Qinling Mountains and adjacent areas, Central China. Geosyst Geoenviron 2(1):100032. https://doi.org/10.1016/j.geogeo.2022.100032

Fan J, Li J, Xia R et al (2014) Assessing the impact of climate change on the habitat distribution of the giant panda in the Qinling Mountains of China. Ecol Model 274:12–20. https://doi.org/10.1016/j.ecolmodel.2013.11.023

Gong K, Huang Z, Qu M et al (2024) Influences of climate change on carbon and water fluxes of the ecosystem in the Qinling Mountains of China. Ecol Indic 166:112504. https://doi.org/10.1016/j.ecolind.2024.112504

He Y, Hu Y, Song J et al (2021) Variation of runoff between southern and northern China and their attribution in the Qinling Mountains, China. Ecol Eng 171:106374. https://doi.org/10.1016/j.ecoleng.2021.106374

He S, Qian H, Liu Y et al (2025a) Water conservation assessment and its influencing factors identification using the InVEST and random forest model in the northern piedmont of the Qinling Mountains. J Hydrol Reg Stud 57:102194. https://doi.org/10.1016/j.ejrh.2025.102194

He D, Qiu K, Yu H et al (2025b) Lithospheric architecture and evolution of the Qinling Orogen of Central China and associated controls on metallogeny. Earth Sci Rev 264:105092. https://doi.org/10.1016/j.earscirev.2025.105092

Jia H, Howard K, Qian H (2020) Use of multiple isotopic and chemical tracers to identify sources of nitrate in shallow groundwaters along the northern slope of the Qinling Mountains, China. Appl Geochem 113:104512. https://doi.org/10.1016/j.apgeochem.2019.104512

Li Y, Qi G, Feng R, Liu K (2022) Responses of radial growth of *Pinus tabuliformis* to climate change at the northern slopes of Qinling Mountains, China. Chin J Appl Ecol 33(8):2043–2050. (in Chinese with English Abstract)

Liu W, Wu J, Xu F, Mu D, Zhang P (2024) Modeling the effects of land use/land cover changes on river runoff using SWAT models: a case study of the Danjiang River source area, China. Environ Res 242:117810. https://doi.org/10.1016/j.envres.2023.117810

Meng Q, Peng X, Zhang S (2024) Spatio-temporal response of vegetation coverage at multiple time scales to extreme climate in the Qinling mountains in Northwest China. Res Cold Arid Re 16(6):302–309. https://doi.org/10.1016/j.rcar.2024.11.005

Peng J, Shen Y, Jin Z et al (2023) Key thoughts on the study of ecogeological environment system in Qinling Mountains. Acta Ecol Sin 43(11):4344–4358. (in Chinese with English Abstract)

Shen Y, Ma T, Peng J et al (2025) Spatial development characteristics of the valley roads and inheritance relationship with water system at the northern foothills of the Qinling Mountains. Acta Geogr Sin 80(1):132–146. (in Chinese with English Abstract)

Shi X, Yang Z, Dong Y, Zhang L (2021) Millennial-scale erosion patterns of the northern Qinling Mountains, Central China: implications for topographical evolution. Geomorphology 382:107670. https://doi.org/10.1016/j.geomorph.2021.107670

Wang H, Duan L, Wang W (2020) Dynamic characteristics and influencing factors of the groundwater level at the Northern Foot of the Qinling Mountains. Northwest Geol 53(2):280–288. (in Chinese with English Abstract)

Wang D, Li P, He X et al (2023) Exploring the response of shallow groundwater to precipitation in the northern piedmont of the Qinling Mountains, China. Urban Clim 47:101379. https://doi.org/10.1016/j.uclim.2022.101379

Xu P, Shen Y, Peng J et al (2024) Typed architecture of valley road in Northern Foothills of Qinling Mountains based on concept of ecological–economic–social collaborative development. Earth Sci 49(12):4564–4575. (in Chinese with English Abstract)

Yu W, Yao T, Lewis SE et al (2014) Stable oxygen isotope differences between the areas to the north and south of Qinling Mountains in China reveal different moisture sources. Int J Climatol 34:1760–1772. https://doi.org/10.1002/joc.3799

Yuan L, Ma T, Han SB et al (2024) Evolution laws and patterns of hydrogeochemistry of groundwater at the Northern Foot of the Qinling Mountains. J Yangtze River Sci Res Inst 41(4):62–69. (in Chinese with English Abstract)

Zhang Z, Duan K, Liu H et al (2022) Runoff projections of the Qinling Mountains and their impact on water demand of Guanzhong region in Northwest China. J Mt Sci 19(8):2272–2285. https://doi.org/10.1007/s11629-022-7358-x

Zhang P, Wu J, Xu F, Liu W (2025) Groundwater resources availability in the Shangzhou Basin Area, China: a modeling research. In: Li P et al (eds) Sustainable groundwater and environment: challenges and solutions. Springer Nature, Cham, pp 237–257. https://doi.org/10.1007/978-3-031-82194-3_11

Chapter 2
Overview of the Study Area

Abstract This chapter provides a comprehensive overview of the northern Qinling Mountains, focusing on its geographical, climatic, topographic, geological, hydrological, and socio-economic characteristics. The chapter details the area's complex climate patterns as well as its distinct geomorphological and geological structures shaped by tectonic evolution. As a key water source region, the northern Qinling contributes significantly to Shaanxi's surface and groundwater resources, yet faces imbalances between water availability and rising demand due to rapid urbanization. The chapter also highlights the region's rich biodiversity, economic growth, tourism potential, agricultural productivity, and mineral and geothermal resource endowments.

Keywords Alluvial plain · Geological structure · Groundwater resources · Surface water · Water management

2.1 Geographical Location

The northern Qinling Mountains are bounded by the Qinling Mountains to the south and the Wei River to the north, with the provincial borders of Shaanxi defining the eastern and western limits. The study area encompasses two primary geographical units: the mountainous region of the northern Qinling and the Guanzhong Basin south of the Wei River. It spans approximately 106°42–110°26′E and 33°42′–34°41′N, covering an area of about 14,600 km^2. The region stretches roughly 336 km in length from east to west, and 10–58 km in width from south to north. Administratively, it includes 19 counties and districts across the cities of Baoji, Xi'an, and Weinan: Tongguan, Huayin, Huazhou, Linwei, Lintong, Lantian, Baqiao, Chang'an, Weiyang, Yanta, Xincheng, Beilin, Huyi, Zhouzhi, Meixian, Taibai, Qishan, Chencang, and Weibin (Fig. 2.1).

P. Li et al., *Water Resources Research in Northern Qinling Mountains, China*, Springer Hydrogeology, https://doi.org/10.1007/978-3-032-09395-0_2

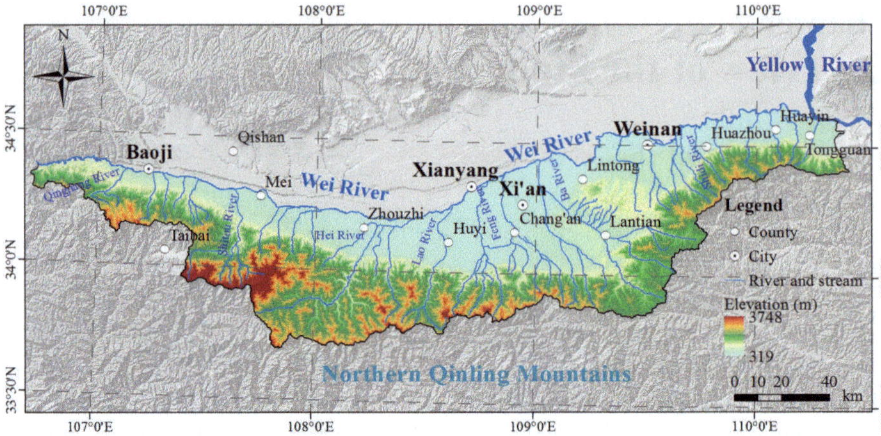

Fig. 2.1 Location map of the northern Qinling Mountains

2.2 Climate Conditions

The northern Qinling Mountains area lies within a warm-temperate semi-arid to semi-humid continental monsoon climate zone. Winter is cold and dry, while summer is hot and rainy. Spring is marked by unstable temperatures and frequent dust storms, whereas autumn is cool and humid, with overcast drizzles and significant diurnal temperature variations. Precipitation is highly seasonal, with 70–80% occurring in summer and early autumn, and only 20–30% in winter and spring (Li et al. 2025). Droughts are common due to high precipitation variability.

The region exhibits pronounced altitudinal climate gradients. The average annual temperature lapse rate is 0.605 °C per 100 m. High-altitude areas above 1700 m are cold and humid, with mean annual temperatures of 6–8 °C; mid- to low-altitude valleys below 1700 m are warmer and wetter, averaging 10–12 °C. Precipitation patterns are influenced by East Asian monsoons and topography, increasing with elevation up to 1000–1400 m (where annual rainfall exceeds 800 mm) before decreasing at higher altitudes.

Rainfall is concentrated in summer and autumn (70–80% of annual totals), with spring and winter being drier (20–30%). Mountainous areas receive 700–900 mm annually, while plains receive 500–700 mm, decreasing from south to north and west to east. The mean annual temperature ranges from 12–14 °C, with recorded extremes of 43.0 °C and −21.9 °C. Key meteorological stations located in the mountainous region report higher rainfall and lower temperatures compared to lowland stations (Fig. 2.2).

Sunshine is abundant (2000–2400 h annually), with a frost-free period of 200–220 days. Spring winds are strongest, reaching 20–40 m/s in high-altitude zones. Annual evaporation ranges from 700 to 1200 mm, peaking in July (Wang et al. 2023). The northeastern Guanzhong Basin has the highest evaporation (>1200 mm), while the western basin and Qinling Mountains record the lowest (<800 mm).

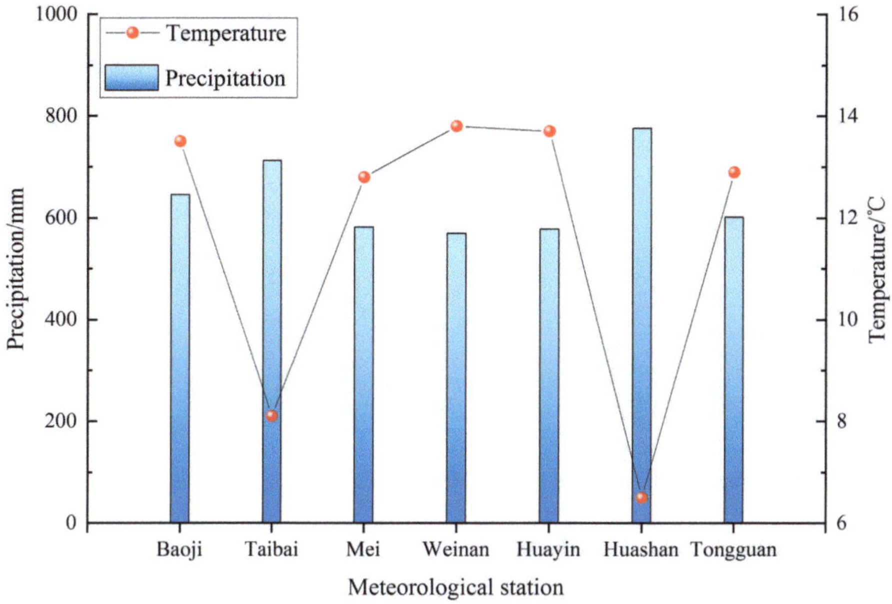

Fig. 2.2 Annual average rainfall and temperature at meteorological stations in the study area

2.3 Topography and Geomorphology

2.3.1 Topography

Elevations in this area range from 319 to 3771 m, sloping from south to north and west to east (Fig. 2.1). The southern Qinling mid-high mountains feature steep, east-west-trending ridges averaging 1000–2320 m, with the highest peak, Taibai Mountain (3771 m), towering 3000 m above the Wei River valley. The northern Guanzhong Basin lies at 325–900 m. Mountainous and hilly areas cover approximately 8600 km^2 (60% of the study area), while plains span about 6000 km^2 (40%).

2.3.2 Geomorphology

The Northern Qinling Mountains area can be divided into alluvial–pluvial plains formed by the Wei River and its tributaries, loess tableland and bedrock mountains in terms of landforms (Fig. 2.3).

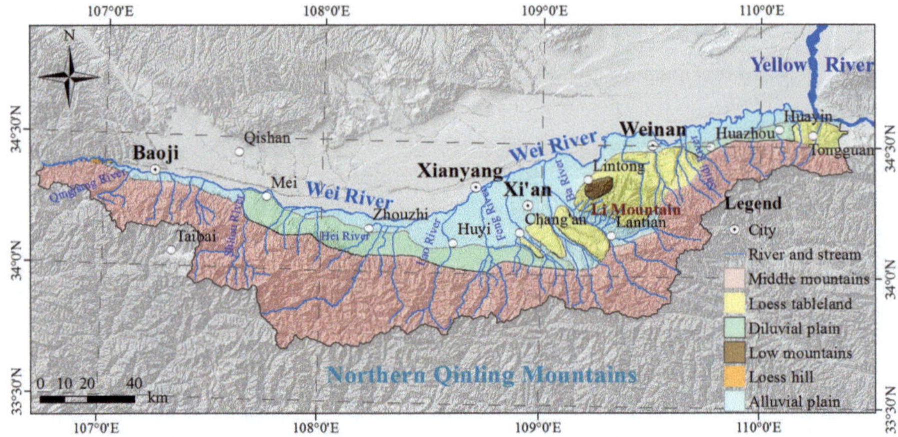

Fig. 2.3 Topography of the northern Qinling Mountains

2.3.2.1 Mountainous Landforms

The mountainous landforms primarily consist of the Qinling mid-to-low mountains and the Li Mountain. The Qinling Mountains stretch along the southern edge of the Guanzhong Basin, with a steep northern slope and a gentler southern slope, higher in the west and lower in the east. They are mainly composed of Proterozoic metamorphic rocks and granites from the Yanshan period.

Many rivers at the northern foot of the Qinling Mountains exhibit a nested valley structure, with a broad U-shaped upper valley and a narrow V-shaped lower valley. Consequently, the river valleys become constricted near the mountain front, while the southern sections tend to widen. This geomorphology results in short, straight rivers with steep valleys, high riverbed gradients, and numerous waterfalls and rapids along the northern Qinling foothills.

The Li Mountain is an intra-basin fault-block mountain within the Guanzhong Basin, composed of Pre-Sinian gneissic granite, quartzite, and granite, overlain by loess. Bounded by normal faults on its northern, western, and southern sides, Li Mountain is a horst-type fault-block mountain with a steep northern slope and a gentler southern slope, peaking at 1302 m. At its northern foothills, alluvial fans are superimposed and extend toward the Wei River, covering its terraces. The southern slope inclines toward the Qinling foothills, characterized by crisscrossing gullies and undulating loess tableland areas.

2.3.2.2 Loess Tablelands

The loess tableland slopes from the Qinling foothills toward the center of the Guanzhong Plain, standing 100–300 m above the Wei riverbed and demarcated by steep cliffs tens of meters high from the river terraces. The loess tableland is mainly distributed in the Xi'an and Weinan regions and can be classified into first-level and second-level loess tablelands.

The first-level loess tableland is continuously distributed over a large area, with abrupt contact against the river valley plains, primarily found in Xi'an, Weinan, and Tongguan. The second-level loess tableland is discontinuously distributed, featured by well-developed gullies and fragmented terrain, sharply contacting the first-level tableland or higher terraces. They are mainly located in Baoji and Lantian.

The Li Mountain loess tableland exhibits significant undulations, forming a wavy loess tableland. In contrast, the Weinan loess tableland has a gentle inclination, presenting as a flat loess tableland. Notably, the section from Huazhou to Huayin is a unique area where the loess tableland is absent (Fig. 2.3).

2.3.2.3 Alluvial–Pluvial Plains

Based on their origin, the plain landforms can be divided into pluvial plains and alluvial plains. The pluvial plains are mainly distributed along the Qinling foothills, consisting of multi-stage pluvial fans. The Qinling piedmont pluvial fans stretch from Mei County–Zhouzhi–Huyi to Huazhou district–Huayin, tilting northward with fine particles and considerable thickness.

The Wei River and its tributary alluvial plains lie in the central part of the basin, formed by sediment deposition from the Wei River and its tributaries, including the Shitou River, Hei River, Feng River, Lao River, and Ba River. The development of river terraces in the Guanzhong Basin is strongly influenced by tectonic activities. Due to the uneven fault-block movements within the basin, the longitudinal and transverse profiles and structural types of the river terraces vary significantly across different sections. West of the Qishan–Mazhao Fault in the Baoji region, where relative uplift occurred earlier and more intensely, six levels of terraces have developed, characterized by bedrock or inset types. East of the Qishan–Mazhao Fault, between Mei County and Xi'an, the Wei River has developed three (locally four) terraces with reduced elevation differences and narrower valleys, featuring inset-type terraces. In Gaoling and Gushi, the Wei River mainly exhibits two terrace levels, classified as superimposed terraces. However, due to the relative uplift of the Lintong Horst and the loess tablelands near Tongguan and Weinan, inset-type third-level terraces reappear in Lintong, Weinan, and Tongguan, while the Gushi Depression between Huazhou District and Huayin only has first-level terraces (Lu et al. 2025).

Overall, the Wei River terraces widen from west to east, with continuous and broad distributions on the northern bank but discontinuous and narrower distributions on the southern bank. The Qinling piedmont pluvial fans are characterized by coarse pluvial sediments with substantial thickness. Buried-type alluvial fans are found in the strongly subsiding areas of Huazhou–Huayin and Chang'an–Zhouzhi, with their fronts gently contacting first- and second-level terraces and their backs showing greater undulations. Additionally, pluvial fans are also well-developed at the back of the loess tablelands along the southern Wei River.

2.4 Geological Settings

2.4.1 Geological Formations

The exposed strata in this region primarily include Archean, Proterozoic, Cambrian, Ordovician, Carboniferous, Permian, Triassic, Cretaceous, Paleogene, and Quaternary formations (Fig. 2.4).

The Archean strata are exposed in areas such as Tongguan and Lantian, predominantly composed of gneiss, including biotite-hornblende gneiss, biotite-plagioclase gneiss, and migmatized biotite-plagioclase gneiss. Proterozoic outcrops in Taibai County, along the Qingjiang River, Tangyu River, and Laoyu River, with lithologies mainly consisting of mica-quartz schist, phyllite, and quartz phyllite.

Cambrian strata are principally distributed in Hanyu and Xiekou, characterized by dolomite, limestone, and marl. Ordovician formations outcrop at Yaowang Mountain and Tangwangling, composed primarily of limestone, marl, chert-banded limestone, shale, and conglomerate. Carboniferous are found near Mei County, Yuhuang Mountain in Taibai County, and Wangjiahe, featured by marl, bauxitic shale, sandy shale, and coal seams.

Permian strata are located in the west of Shichuan River such as areas in Huangtian Town, Kou Town, and Mujianghekou, with upper sections comprising sandstone, mudstone, and shale, while lower sections containing coal-bearing deposits. Triassic formations are distributed along the Jing River, Baliping, Weiziping, and Mujianghekou, consisting mainly of monzogranite, sandstone, sandy shale, and shale.

Cretaceous strata occur along the Xi River and Pantao River north of Fengxiang County, the Jinling River in Long County, and near Baiyun Town in Taibai County, predominantly composed of sandstone and conglomerate. The Paleogene–Neogene formations are exposed in Bailu Tableland and Li Mountain areas, featured by

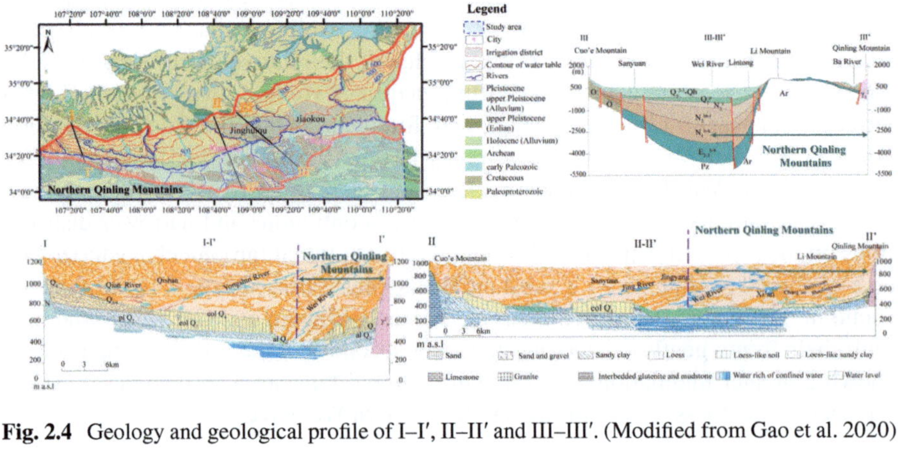

Fig. 2.4 Geology and geological profile of I–I′, II–II′ and III–III′. (Modified from Gao et al. 2020)

sandstone, siltstone, and mudstone. Quaternary deposits are extensively distributed throughout the region, primarily consisting of loess with occasional gravel and pebble layers.

2.4.2 Geological Structures

The study area exhibits complex geological structures with intense folding and well-developed fault systems. Through prolonged tectonic evolution, the northern margin of the Qinling orogenic belt displays significant compressional deformation features, manifested as compressional schistosity zones, cataclasite, mylonite, and fault structures.

The Northern Qinling Piedmont Fault Zone extends approximately 200 km from Baoji in the west to Lantian in the east. Tectonically positioned at the junction between the Qinling Fold System and the Sino-Korean Paraplatform (North China Craton), this fault zone precisely marks the transitional boundary between the Qinling orogenic belt and the southern margin of the Wei River Basin, serving as the natural divide between the Qinling Mountains and the Wei River Basin (Fig. 2.4).

West of Hei River mouth in Zhouzhi County, the fault extends in a North–West–West (NWW) direction, and gradually shifts to an approximate East–West (EW) orientation eastward. The fault plane is near-vertical with a maximum northward dip angle of about 70°, representing a normal fault where the southern block forms the hanging wall and the northern block the footwall. The area features well-developed fractures and joints, with intense rock weathering resulting in poor structural integrity. Throughout its geological evolution, the Northern Qinling Piedmont Fault Zone has undergone multiple phases of displacement, deformation, convergence, and extension processes.

2.5 Water Resources

The northern Qinling Mountains serve as the headwaters for the Wei River, the biggest tributary of the Yellow River, and its numerous tributaries. The Wei River hydrological system covers 24% of the Qinling mountainous area and is the primary water source for the Guanzhong urban agglomeration. Water resources in this region are critical for the socioeconomic development of the Guanzhong area (Li et al. 2025).

Within Shaanxi Province, the Qinling Mountains hold approximately 22.2 billion m^3 of water resources, accounting for 50% of the total water supply in this province, making it the key water conservation zone in Shaanxi. Specifically, the northern Qinling foothills contribute 4 billion m^3 of water, representing 51% of the surface water resources in the Guanzhong region. This area is not only a vital water retention zone for the Wei River Basin, but also the lifeline of Guanzhong's aquatic ecosystem, sustaining ecological water flows, economic sustainability, and urban water demands. Despite occupying only 26% of Guanzhong's land area (7% of Shaanxi's total), the northern Qinling foothills support nearly 40% of Guanzhong's population (26% of Shaanxi's total) and 51% of Guanzhong's GDP (32% of Shaanxi's total), making it the lifeline and core of the entire Guanzhong ecosystem.

Regarding water resources utilization, the terrace regions have the highest level of exploitation over the Guanzhong Basin, followed by the loess tablelands, while mountainous areas have the lowest level of water resources development. The per capita water resources in the Wei River Basin are equivalent to only 12.4% of the national average. Cities including Baoji, Xianyang, Xi'an, and Weinan all experienced overexploitation of groundwater in the past. With accelerating urbanization, water demand continues to increase, and the supply-demand conflicts will become more prominent.

2.5.1 Precipitation

Precipitation in the northern Qinling Mountains shows distinct spatial and temporal distribution characteristics. In general, precipitation is higher in the south than the north, and higher in summer and autumn than in spring and winter. As illustrated in Fig. 2.5, the mean annual precipitation ranged from over 550 mm to less than 850 mm, with most areas receiving 600–800 mm. The highest precipitation (>850 mm) occurred at the Qinling watershed divide, while the lowest precipitation (<600 mm) was observed in the Weinan City and Xi'an City near the Wei River. By administrative division, Xi'an had the highest precipitation in 2023 at 1016.0 mm (multi-year average: 729.3 mm), followed by Baoji with 751.5 mm (multi-year

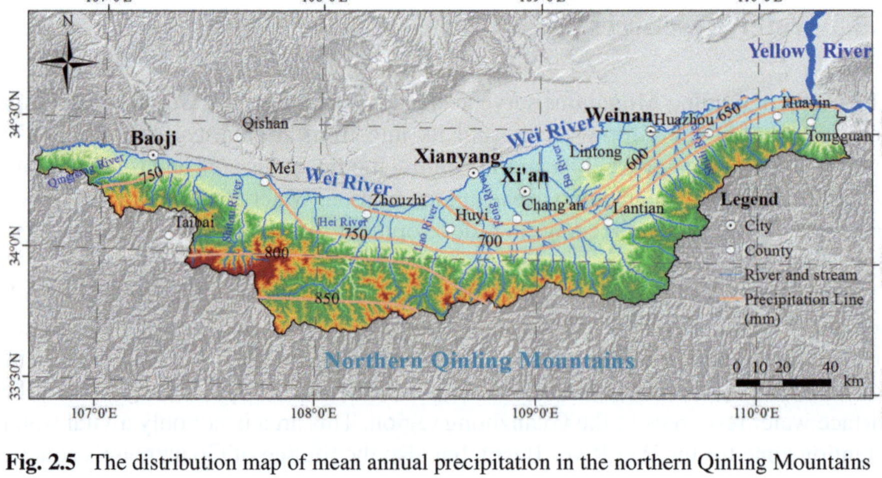

Fig. 2.5 The distribution map of mean annual precipitation in the northern Qinling Mountains

average: 679.9 mm). Weinan had the least precipitation at 731.4 mm (multi-year average: 588.7 mm).

2.5.2 Surface Water

The northern piedmont of the Qinling Mountains belongs to the Yellow River Basin, characterized by numerous rivers and valleys. There are more than 100 major valleys above the mouths of the gullies, eventually converging into 34 rivers that flow into the Wei River, serving as an important water resource for domestic and industrial uses in the Guanzhong region (He et al. 2025). The rivers on the northern piedmont of the Qinling Mountains generally range in length from 10 to 100 km, with catchment area typically between 10 and 500 km^2. Among them, there are 27 primary and secondary tributaries with catchment area of 10–100 km^2, such as Weijia Gully, Mawei River, and Xisha River. There are 21 primary and secondary tributaries with catchment area of 100–500 km^2, such as Qingjiang River, Qingshui River, Tangyu River, Ling River, and Luowen River. Additionally, there are 5 primary and secondary tributaries with catchment area exceeding 500 km^2, such as Shitou River, Hei River, and Ba River.

The water resources within the Qinling catchment exhibit significant spatial and temporal variability, with large interannual and intra-annual fluctuations. The rainy season is prone to flooding, while the dry season often leads to drought and water shortages. Runoff is primarily concentrated between July and October, with marked fluctuations throughout the year. The study area encompasses three cities, Xi'an, Baoji, and Weinan, which are densely populated and have a high demand for water resources. However, the streams are short and the headwaters run fast, making water resource development challenging. There is a pronounced imbalance between supply and demand, and water resources have become a critical constraint on the socio-economic development of the region.

In 2023, the total surface water resources of Shaanxi Province amounted to 50.869 billion m^3, equivalent to a runoff depth of 247.4 mm. By administrative division, Xi'an had the highest runoff depth in 2023, reaching 362.1 mm, which is much higher than the provincial average. Baoji had a runoff depth of 188.5 mm, while Weinan recorded the lowest runoff depth at 103.4 mm. The runoff depths of Baoji and Weinan were both below the provincial average (Shaanxi Province Department of Water Resources 2024).

The hydrochemical types of the rivers in the northern piedmont of the Qinling Mountains are mostly calcium bicarbonate type and sodium bicarbonate type. The total dissolved solids (TDS) generally ranges between 100 and 500 mg/L, which is classified as fresh water (Nsabimana et al. 2023). The total hardness (TH) is lower in the western rivers, moderate in the central rivers, and highest in the eastern rivers. The water quality in various catchments shows considerable fluctuation. During the wet season, the main rivers in the northern piedmont of the Qinling Mountains can generally achieve better water quality than the dry season. Additionally, river sections downstream of the valley mouths often experience flow interruptions during the dry season, and some rivers even remain dry throughout the year. Surface water in the northern Qinling Mountains has superior quality to that in the north of the Wei River Plain, which is unsuitable for irrigation and industrial purposes (Nsabimana et al. 2023). However, The direct discharge of domestic sewage and industrial wastewater, along with pollution from agriculture and animal husbandry, poses a severe threat to the aquatic ecosystems of the region. Furthermore, the northern piedmont of the Qinling Mountains also faces ecological challenges such as wetland shrinkage, deficits in water quantity in rivers and lakes, and the reduction in lake and wetland areas.

2.5.3 Groundwater

The multi-year average groundwater resources in the northern piedmont of the Qinling Mountains are approximately 2 billion m^3. In the plains area of the study region, the recharge of groundwater comes mainly from infiltration of precipitation, infiltration of surface water, and lateral recharge from the mountainous areas (Nsabimana et al. 2025). In the hilly areas, groundwater resources are mainly recharged by precipitation infiltration. Precipitation infiltration and river leakage are the primary recharge sources for groundwater in the plains area of the study region, accounting for about 40% of the total groundwater recharge. Groundwater is well developed in the study region, far exceeding the national average level.

In 2023, the groundwater resources in the three cities (Baoji City, Xi'an City, Weinan City) amounted to 4.456 billion m^3. By administrative division, Xi'an City had the highest groundwater resources, at 1.555 billion m^3, followed by Baoji City with 1.548 billion m^3, while Weinan City had the least, at 1.353 billion m^3. In the Guanzhong Basin, the area where the groundwater level depth exceeds 40 m is the largest, accounting for about 24.32% of the basin area. Areas with groundwater level depths of 20–40 m, 8–20 m, 4–8 m, 2–4 m, and less than 2 m accounted for 14.28%, 12.05%, 14.70%, and 14.75% of the basin area, respectively. In recent years, the groundwater level in the northern Qinling Mountains has been continuously rising (Shaanxi Province Department of Water Resources 2024).

2.6 Population, Economy, and Resources

In 2023, the study area had a permanent population of approximately 13.5 million. Based on administrative divisions, the population distribution from highest to lowest was as follows: Yanta District, Chang'an District, Weiyang District, Baqiao District, Lianhu District, Linwei District, Beilin District, Lintong District, Xincheng District, Huyi District, Zhouzhi County, Weibin District, Lantian County, Mei County, Huazhou District, Huayin City, and Tongguan County (Table 2.1, Shaanxi

Table 2.1 Permanent population in different districts and counties in the study area (Shaanxi Provincial Bureau of Statistics 2024)

County/district	Population	County/district	Population	County/district	Population
Tongguan	123,100	Huayin	201,300	Huazhou	263,700
Linwei	904,400	Lintong	687,500	Baqiao	1,066,900
Chang'an	1,644,200	Yanta	2,129,100	Beilin	778,600
Xincheng	627,000	Weiyang	1,625,700	Lianhu	1,037,500
Lantian	494,500	Huyi	595,200	Zhouzhi	558,800
Mei	275,300	Weibin	543,200		

Provincial Bureau of Statistics 2024). Among these, Yanta District recorded the highest population at 2.13 million, accounting for 15.71% of the total population. This was followed by Chang'an District (1.64 million; 12.13%), Weiyang District (1.63 million; 11.99%), Baqiao District (1.07 million; 7.87%), and Lianhu District (1.04 million; 7.65%).

The study area has experienced sustained economic growth over the years. In 2023, the region's Gross Domestic Product (GDP) reached 1212.411 billion Chinese yuan. GDP by administrative division is detailed in Table 2.2. Among these, Yanta District had the highest GDP at 311.894 billion yuan, followed by Weiyang District (156.545 billion yuan) and Chang'an District (145.517 billion yuan). In terms of year-on-year growth rates, Chang'an District, Zhouzhi County, and Huyi District, exhibited the most significant increases, with growth rates of 8.0%, 7.9%, and 7.4%, respectively (Shaanxi Provincial Bureau of Statistics 2024).

The northern Qinling Mountains are rich in tourism resources, including a wealth of historical relics, cultural heritage sites, and intangible cultural heritage. The main valleys along the mountain front within the study area possess excellent natural landscapes characterized by scenic beauty and diverse forms, making them ideal for tourism development. Notable tourist attractions in the region include Mount Hua, Mount Taibai, Louguantai, the Qinling National Botanical Garden, Honghe Valley, Zhonglvping, and the Bailuyuan Film and Television City.

In addition to cultural and scenic resources, the study area is home to a rich diversity of wildlife, including many rare and protected species. These include the giant panda, golden snub-nosed monkey, golden takin, clouded leopard, serow, red panda, leopard cat, golden pheasant, oriole, white crane, hwamei (*Garrulax canorus*), mallard, egret, and dove. The ecological conditions of the northern Qinling region provide an excellent natural habitat for wildlife conservation and biodiversity.

Due to its fertile soils and abundant hydrothermal conditions, the study area is highly suitable for agricultural development. In the loess tableland regions, where the terrain is relatively flat, economic forests and various crops are widely cultivated. In the low alluvial fan areas, where the elevation gradually rises from north to south, economic forests and other agricultural crops are also commonly planted.

The northern Qinling Mountains represent the earliest and most extensively developed region for commercial kiwifruit cultivation in China. It remains the

Table 2.2 Gross Domestic Product (GDP) in different districts and counties of the study area (Shaanxi Provincial Bureau of Statistics 2024)

County/district	GDP (billion yuan)	Growth rate (%)	County/district	GDP (billion yuan)	Growth rate (%)	County/district	GDP (billion yuan)	Growth rate (%)
Tongguan	4.427	1.0	Huayin	7.968	1.6	Huazhou	10.255	0.7
Linwei	50.034	3.9	Lintong	27.501	4.4	Baqiao	70.961	5.4
Chang'an	145.517	8.0	Yanta	311.894	4.6	Beilin	114.570	4.3
Xincheng	72.210	5.7	Weiyang	156.545	0.8	Lianhu	84.644	5.4
Lantian	16.129	5.2	Huyi	40.639	7.4	Zhouzhi	18.406	7.9
Mei	17.703	1.9	Weibin	63.008	3.9			

country's largest and most industrialized area for kiwifruit production, leading the nation in both cultivation area and output. Beyond cultivated crops, the region is also rich in wild plant resources, including ferns, fungi, mosses, hardwoods, shrubs, and a variety of wild medicinal herbs. Furthermore, the study area holds abundant mineral resources. Metallic minerals such as iron, copper, gold, and silver are found here, alongside substantial reserves of non-metallic minerals, including marble, granite, gravel, serpentine, emerald, and clay.

In addition, many parts of the study area exhibit favorable conditions for geothermal resource formation. As a result, the region possesses significant geothermal energy potential, with promising prospects for future development and utilization.

References

Gao Y, Qian H, Huo C et al (2020) Assessing natural background levels in shallow groundwater in a large semiarid drainage basin. J Hydrol 584:124638. https://doi.org/10.1016/j.jhydrol.2020.124638

He S, Qian H, Liu Y et al (2025) Water conservation assessment and its influencing factors identification using the InVEST and random forest model in the northern piedmont of the Qinling Mountains. J Hydrol Reg Stud 57:102194. https://doi.org/10.1016/j.ejrh.2025.102194

Li X, Li P, Liu J (2025) Water security in the Qinling Mountains: current status, climate and human influences, and adaptive strategies. Int J Water Resour Dev. https://doi.org/10.1080/07900627.2025.2478387

Lu H, Li P, Wang D, Xu F, Fida M (2025) Hydrochemical characteristics and transformation relationship of surface water and groundwater in the Hua County, Guanzhong Plain (China). Water Environ Res 97(6):e70115. https://doi.org/10.1002/wer.70115

Nsabimana A, Li P, Alam SMK et al (2023) Surface water quality for irrigation and industrial purposes: a comparison between the south and north sides of the Wei River Plain (northwest China). Environ Monit Assess 195:696. https://doi.org/10.1007/s10661-023-11263-0

Nsabimana A, Li P, Alam SMK, Fida M (2025) Assessing phreatic and confined water quality in the Guanzhong Basin, China: a novel integrated-weight water quality index approach. Environ Earth Sci 84(10):260. https://doi.org/10.1007/s12665-025-12249-6

Shaanxi Province Department of Water Resources (2024) Shaanxi Water Resources Bulletin 2023. https://slt.shaanxi.gov.cn/zfxxgk/fdzdgknr/zdgz/szygb/202411/t20241119_3197876.html. Accessed 18 July 2025

Shaanxi Provincial Bureau of Statistics (2024) Shaanxi Statistical Yearbook 2024. China Statistics Press, Beijing. https://tjj.shaanxi.gov.cn/tjsj/ndsj/tjnj/sxtjnj/index.html?2024. Accessed 18 July 2025

Wang D, Li P, He X et al (2023) Exploring the response of shallow groundwater to precipitation in the northern piedmont of the Qinling Mountains, China. Urban Clim 47:101379. https://doi.org/10.1016/j.uclim.2022.101379

Chapter 3
Main Rivers Originating from the Northern Qinling Mountains

Abstract This chapter provides a detailed examination of the main rivers originating from the northern Qinling Mountains, which serve as critical water sources for the Guanzhong Plain. A total of 253 river valleys span the northern slopes, with rivers classified into two major types based on geomorphology and recharge sources, narrow mountainous rivers fed primarily by precipitation and broader valley rivers sustained by groundwater. The chapter provides case studies on key rivers across Baoji, Xi'an, and Weinan. Each case includes hydrological characteristics, water quality conditions, recharge mechanisms, ecological functions, and infrastructure projects such as reservoirs. The chapter emphasizes the importance of these rivers for urban water supply, ecological conservation, and regional water security, laying the foundation for further hydrological assessment and integrated water resource management strategies.

Keywords Major revers · Water resources · Surface water · Water management · Ecosystem

3.1 Overview of the Rivers Originating from the Northern Qinling Mountains

There are as many as 253 river valleys along the northern Qinling Mountains, which not only facilitate transport between the northern and southern Qinling regions, but also supply a steady stream of water to the Guanzhong Plain (Ren et al. 2024; Nsabimana et al. 2023). The major rivers are shown in Fig. 3.1 and the main river characteristics on the northern Qinling Mountains are detailed in Table 3.1. Due to the influence of the Qinling Mountains' terrain, rivers along the northern Qinling can generally be categorized into two types: those with narrow valleys and those with broad and flat river valleys. The former are typically found within steep mountainous gullies, while the latter are located where the mountains transition into the plains. These two types also differ in their sources of recharge. Rivers in steep

P. Li et al., *Water Resources Research in Northern Qinling Mountains, China*, Springer Hydrogeology, https://doi.org/10.1007/978-3-032-09395-0_3

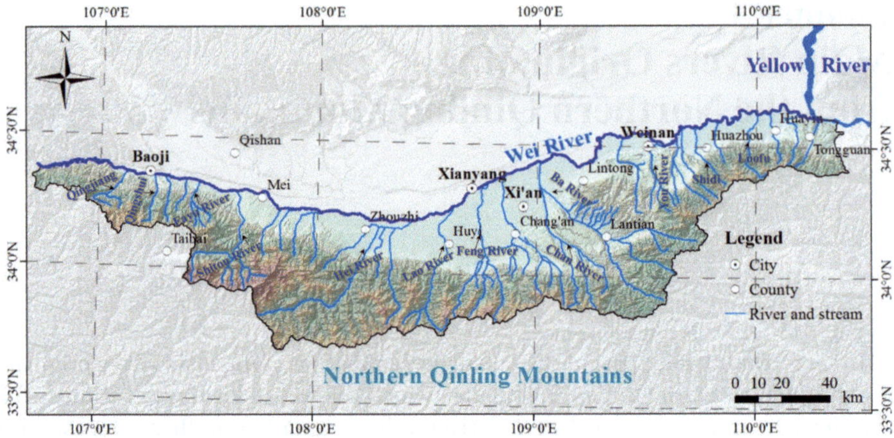

Fig. 3.1 Main rivers in the northern Qinling Mountains

Table 3.1 Characteristics of the main rivers originating from the northern Qinling Mountains

City	Rivers	Catchment area (km^2)	Length (km)	Rivers	Catchment area (km^2)	Length (km)
Xi'an	Hei River	2258	125.8	Zao River	315.5	33.5
	Gengyu River	81.5	38.2	Ling River	292	53.9
	Xin River	303.8	40	Xi River	117.7	33.2
	Lao River	663	77.7	Yuchuan River	27.2	12.5
	Feng River	1386	78	Ba River	2581	109
Baoji	Shitou River	778	68.6	Mawei River	85.4	26.9
	Tangyu River	386	43.9	Panxi River	67.5	25.5
	Tongyu River	27	17.8	Fayu River	155.1	26.3
	Maili River	63.7	18.2	Bawang River	177.1	34.2
	Qingjiang River	234	43	Xisha River	67	25.8
	Qingshui River	162.5	28.4	Dongsha River	118	30.3
Weinan	You River	269.5	41.5	Luofu River	140	47.2
	Chishui River	300	40.2	Liuye River	134.9	30.6
	Yuxian River	141.5	41.1	Changjian River	118.6	29.4
	Luowen River	115.1	33.8	Bailongjian River	119.4	26.6
	Shidi River	188.7	36.8	Shuangqiao River	331	101.6
	Fangshan River	11.9	9.9	Tong River	115.4	24.1

mountainous zones are primarily fed by precipitation, whereas those in the transitional zones are mainly sustained by groundwater.

The total annual surface water resources in the northern Qinling Mountains average approximately 4 billion m^3, accounting for 61% of the Wei River's surface water resources. This makes the region a crucial water source for industrial and agricultural production, as well as for domestic use in the Guanzhong Plain. Notably, reservoirs such as the Shitou River Reservoir and the Hei River Reservoir serve as major water supply sources for residential and industrial consumption in Xi'an City. These rivers are primarily recharged by snowmelt and spring water, which are generally clear, unpolluted, and highly transparent. There are six rivers in the region with annual runoff exceeding 100 million m^3: Qingjiang River, Shitou River, Laoyu River, Feng River, Hei River, and Ba River. Due to intense human activities in the areas along the Wei River, surface water is polluted, and river water pollution by microplastics, trace metals and nitrate have been well reported (Zhao et al. 2023; Lu et al. 2025; Li et al. 2024).

3.2 Main Rivers in Baoji City

3.2.1 Qingjiang River

The Qingjiang River, formerly known as Jiang River, originates from the northern slope of Yuhuang Mountain. It is a tributary on the right bank of the Wei River and serves as the primary drinking water source for Baoji City. The main channel of the river is 43 km long, with a gradient of 31.8‰, a total elevation drop of 2032 m, and a drainage area of 234.4 km^2, characterized by steep rock walls and deep valleys along its riverbanks. The upper reaches of the Qingjiang River, also known as the Shensha River, flow from southeast to northwest, turn northeast at Qingshiya, and pass through Guanyinya, Yangjiawan, and Yimenpu. At Erliguan, it receives the right-bank tributary Lijia River, and joins the Wei River at Shijiaying in Weibin District. The basin is located on the northern slope of the western segment of the Qinling fold belt, with well-preserved vegetation coverage of approximately 80%, favorable for water conservation, and relatively stable runoff during dry seasons.

According to surveys, the upper reaches of the Qingjiang River feature large boulders on both banks with high roundness, while the riverbed is dominated by smaller stones with poor sorting. The water is relatively clear, with visible fish. In the middle reaches, granite is present on both banks, vegetation is lush, and large rocks emerge in the river center. The river flows fast with low sediment content, the terrain slopes from west to east, and the water remains clear. However, most of the downstream channel is mostly dried up, with clear and shallow water flowing slowly in local sections. No visible litter is present, indicating the effective implementation of river regulation measures in the downstream section of the Qingjiang River since 2016. Figure 3.2 displays photographs taken from the sites of upstream, midstream, and downstream.

Fig. 3.2 Qingjiang River (upstream, midstream and downstream from left to right)

3.2.2 *Qingshui River*

The Qingshui River, a primary right-bank tributary of the Wei River, derives its name from its clear and transparent water (Qingshui means clear water in Chinese). Qingshui River originates from two sources on the northern slope of the Qinling Mountains: one on the northern side of Jifeng Mountain and the other at the foot of Mo'ao Mountain. These two water sources merge to form the Xicha River, which then joins with the Bayu River before flowing into the Wei River. The main stem of the Qingshui River is 28.9 km long, with a drainage area of 1608 km^2 and an annual runoff volume of 66.625 million m^3. The Qingshui River has two major tributaries: the Dongcha River and the Xicha River. The Dongcha River serves as the main stem and includes tributaries such as the Daqing River, Xiaoqing Lake, Xiaogangou, Liangzigou, Maijiagou, and Daohuigou. The Xicha River includes tributaries like Donggou and Qinghegou.

Surveys indicate the presence of a factory and a wastewater discharge outlet to the west of where the Qingshui River exits the mountain. The river emits a pungent odor, although the water appears relatively clear. Algal blooms are common, and the riverbed is dominated by poorly sorted stones. In the downstream section of the Dongcha River, the water is turbid and the riverbed is slightly silted, likely due to nearby construction activities. The Xicha River exhibits better water quality than the Dongcha River, with crystal-clear water and visible fish. After exiting the Qinling Mountains, numerous small industrial facilities are distributed along both banks of the Qingshui River, with industrial pollution constituting the primary environmental threat. Figure 3.3 shows the on-site photographic documentation.

3.2.3 *Fayu River*

The Fayu River was historically known as Fanxi. The Fayu River is a primary right-bank tributary of the Wei River, with its source located on the northern slope of Qingfeng Mountain in Tianwang Town, Chencang District, at the northern foot of the Qinling Mountains. The main stem of the Fayu River is 27.1 km long, with an average channel gradient of 31.8‰ and a multi-year mean runoff of 81.37 million

Fig. 3.3 Qingshui River (upstream, midstream and downstream from left to right)

Fig. 3.4 Fayu River (upstream reservoir, midstream and downstream from left to right)

m^3. Among its several tributaries, the most prominent are the Wupan, Shibapan, Shierpan, and Shian Rivers. The total drainage area is 156.0 km^2, and it joins the Wei River at Xujia Village, Tianwang Town.

Field investigation shows that the upper reaches of the Fayu River contain large amounts of gravel and stones, and the water is clear and slow-moving, with moss-covered stones and a riverbed dominated by fine sand in relatively flat terrain. The middle reaches have slightly undulating terrain, the water remains relatively clear, and the flow is slow. Construction activities near the lower reaches have caused water turbidity and the accumulation of construction debris and wastes. The flow rate in the lower reaches is moderate, the riverbed is composed mainly of sand and gravel, and moss is sparsely distributed on the stones. Its tributary, the Shierpan River, features steep terrain and a winding river course, with crystal-clear water and a sandy riverbed. The surrounding mountains are steep and densely forested, offering scenic views that attract a continuous stream of tourists for recreational activities. Field photographs are shown in Fig. 3.4.

3.2.4 Shitou River

The Shitou River originates from Taibai Mountain, the main peak of the Qinling Mountains, and was historically known as Wugongshui or Xieshui. It is a first-order tributary of the Wei River. The main channel extends 68.6 km, with abundant flow

Fig. 3.5 Shitou River (upstream reservoir, midstream and downstream from left to right)

and consistently high water quality that meets Class I surface water standards (excellent water quality). It serves as a high-quality drinking water source for Guanzhong cities such as Xi'an, Xianyang, Baoji, and Yangling. In 2016, it was listed as a state key drinking water source and is now honored as the "Water Tower of Guanzhong".

The Shitou River Reservoir is situated at the upstream of the Shitou River. The dam stands 114 m high, with a total storage capacity of 147 million m^3, an effective capacity of 120 million m^3, and an annual regulation capacity of 270 million m^3. Construction of the reservoir began in 1971, and impoundment commenced in 1980. It is a large-scale (Class II) multipurpose hydraulic project that is used for flood control, urban water supply, hydropower generation, and agricultural irrigation. Shaanxi Province is characterized by structural water scarcity. The Shitou River Reservoir has drawn significant attention due to its ecological integrity, high water quality, and strategic location for supplying water to towns and cities in the Guanzhong Plain. The Xi'an water supply project was completed in June 1996, followed by the Xianyang project in October 2009, the Yangling project in October 2012, and the Baoji project in August 2013. This has formed a regional urban water supply network centered around the Wei River and anchored by cities such as Xi'an, Xianyang, Yangling, and Baoji, covering the central and western Guanzhong area, with an annual supply stabilized at approximately 100 million m^3. Field photographs showing the upstream reservoir, midstream and downstream of the Shitou River are shown in Fig. 3.5.

3.3 Main Rivers in Xi'an City

3.3.1 Hei River

The Hei River Basin is centered around the Hei River as the main stem, with tributaries including the Sha River, Tianyu River, Gengyu River, Niyu River, and Xiluoyu River. The basin exhibits a feather-shaped drainage pattern, with tributaries running parallel along both banks of the Hei River and converging in the plains. The total length of the Hei River is 125.8 km, covering an area of 2258 km^2, with an annual

Fig. 3.6 Hei River (upstream reservoir, midstream and downstream from left to right)

runoff depth of 362 mm. The water quality of the Hei River is generally good, with an average annual runoff of 817 million m^3. The maximum flood discharge recorded at the Heiyukou Hydrological Station in 1980 was 3040 m^3/s.

The Jinpen Reservoir is located in the upstream of the Hei River, and lies 86 km from Xi'an. It is a large (Class II) multipurpose hydraulic project primarily for urban water supply, with additional functions including irrigation, hydropower generation, and flood control. It also serves as a major water source for Xi'an. Jinpen Reservoir is a critical component of the Hei River Diversion Project, with construction starting in 1996 and completion achieved by the end of 2002 (Meng and Xu 2024). The reservoir complex consists of key structures such as a dam, spillway tunnel, flood discharge tunnel, diversion tunnel, downstream power station, and seepage control works for the paleo-channel. The dam site controls a catchment area of 1481 km^2, with a long-term average annual runoff of 667 million m^3. The reservoir has a total storage capacity of 200 million m^3 and an annual regulation capacity of 428 million m^3. Currently, Jinpen Reservoir also functions as a regulating reservoir for the Han-to-Wei River Water Diversion Project, making it the largest storage reservoir in Xi'an, with a daily water supply capacity of 1.1 million tons, effectively meeting the domestic and industrial water demands of the main urban area. Figure 3.6 shows the on-site photographic documentation.

3.3.2 Lao River

The Lao River, one of rivers forming the "Eight Waters Encircling Chang'an", originates from the northern Qinling Mountains and flows through the Huyi District of Xi'an. The Lao River, a primary tributary of the Wei River, extends 77.7 km, comprising 45.7 km of mountainous stretches and 32 km across plains. It drains a basin of 663 km^2, with an average riverbed gradient of 9.5‰. The upper reaches of the Lao River are characterized by a V-shaped gorge. Narrow and deep valleys dominate the middle and upper sections, whereas the downstream areas broaden progressively. The Lao River exhibits an asymmetric drainage pattern, with the majority of its principal tributaries concentrated on the left bank. Major tributaries of the Lao River include the Gan River, Zaoyu River, Liyu River, and Baima River, among which the Gan River is the largest. Within the mountainous region, the Gan River

Fig. 3.7 Lao River (upstream, midstream and downstream from left to right)

stretches for 17 km, drains an area of 69.1 km^2, and yields an annual runoff of 24.18 million m^3. Figure 3.7 shows the on-site photographic documentation of upstream, midstream, and downstream of the Lao River.

3.3.3 *Feng River*

Originating in Fengyu valley on the northern slope of the Qinling Mountains, the Feng River is a right-bank tributary of the Wei River, receiving contributions from the Jue River, Taiping River, and Gaoguan River. Traversing Chang'an District and Huyi District, the river extends 78 km with a basin area of 1386 km^2, an average channel gradient of 8.2‰, and a multi-year mean annual runoff of 429 million m^3. The Feng River Basin is characterized by a warm-temperate, semi-arid to semi-humid continental monsoon climate. The mean annual precipitation in the basin is 824.4 mm, yielding total water resources of 520 million m^3, comprising 400 million m^3 of surface water and 240 million m^3 of groundwater. Land use in the basin comprises farmland, forest, grassland, water bodies, urban areas, and rural settlements, with forest, cropland, and grassland constituting the largest shares at 41.6%, 36%, and 16%, respectively (Zhi et al. 2016). Forest and grassland dominate the upstream region, whereas the midstream and downstream areas are primarily characterized by agricultural land, urban development, and rural settlements. Agricultural practices involve intensive use of chemical fertilizers, with nitrogen-based fertilizers comprising nearly 50% of total fertilizer input. Figure 3.8 shows the different characteristics of upstream, midstream and downstream of the Feng River.

3.3.4 *Ba River*

The Ba River originates in the Jiudaogou region, situated on the south of the Jianyuling Mountain at the border of Lantian County and Huazhou District. Among the 8 rivers forming the historic "Eight Waters Encircling Chang'an", the Ba River carries the highest flow and has the best water quality. It is a primary tributary of the

Fig. 3.8 Feng River (upstream, midstream and downstream from left to right)

Fig. 3.9 Ba River (upstream, midstream and downstream from left to right)

Wei River, extending 109 km in length with a basin area of 2581 km^2 and an average bed gradient of 6.0‰. The multi-year average runoff of the Ba River is 21.81 m^3/s, with an annual runoff amount of 688 million m^3. The Ba River receives recharge from five tributaries, with the Chan River being the largest, which covers 41.6% of the entire watershed. Figure 3.9 shows the different characteristics of upstream, midstream and downstream of the Ba River.

The climate of the Ba River basin is classified as warm-temperate semi-humid continental monsoon, characterized by pronounced seasonal changes in both temperature and moisture. Due to climatic and topographical factors, precipitation across the Ba River basin exhibits significant spatiotemporal variability. Rainfall predominantly occurs between July and September, with greater precipitation generally occurring in the southern regions than in the northern parts of the basin. According to meteorological records (2005–2019) from Lantian County, average annual precipitation within the basin ranged from 502 to 873 mm. Average annual temperature in the basin ranged from 13.0 to 14.8 °C. Historical records showed peak summer temperatures surpassing 45 °C and winter temperatures dropping below −20 °C.

Based on groundwater depth and hydrogeological properties, groundwater is categorized into two aquifer systems: the phreatic aquifer and confined aquifer, generally separated by an impermeable interface located at approximately 70 m below the ground surface. The phreatic aquifer primarily consists of Quaternary alluvial deposits, comprising loess, loess-like soils, silty clay, sand, and gravel. The base of this aquifer extends to depths of approximately 70–80 m below the ground surface, with groundwater levels ranging from 10 to 30 m and a hydraulic conductivity between 10 and 34 m/day. The

Fig. 3.10 Chan River (upstream, midstream and downstream from left to right)

aquifer is recharged by both precipitation and river infiltration, while agricultural abstraction is the primary discharge pathway (Zhao et al. 2024).

3.3.5 Chan River

Chan River, recognized as one of rivers forming the "Eight Waters Encircling Chang'an", constitutes the largest primary tributary of the Ba River. The Chan River originates from Ziyun Mountain at the northern Qinling Mountains, located in Tangyu Town, Lantian County. Flowing from south to north, it joins the Ba River in Baqiao District. The river spans 64 km in length, covers a watershed area of 760 km^2, and features three dendritically distributed tributaries. Characterized by a warm temperate continental monsoon climate, the basin receives an annual average rainfall of about 600 mm, primarily concentrated from June to September. The Chan River's average annual runoff in Yanta District is 175 million m^3, with a maximum flood peak of 632.5 m^3/s. The runoff fluctuates greatly between the dry and wet seasons, resulting in a very low annual utilization rate of water resources. The river maintains an annual average discharge of 175 million m^3 in Baqiao District, where it also carries approximately 2.5 million tons of sediment annually. The characteristics of upstream, midstream and downstream of the Chan River are shown in Fig. 3.10.

3.4 Main Rivers in Weinan City

3.4.1 You River

The You River, a tributary of the Wei River, originates from Erlang Mountain on the northern Qinling Mountains and flows through Weinan City. It is formed by the confluence of the Qingshui River (this river has the same river name with that in Baoji City) in the east and the Choushui River in the west. The total length of the

Fig. 3.11 You River (upstream, midstream reservoir and downstream from left to right)

river is 41.5 km, with a basin area of 269.5 km^2. In the Weinan City, major water bodies associated with the You River include the You River Reservoir, irrigated farmlands, and the downstream Yu River Wetland Park (Fig. 3.11). The multi-year average runoff of the You River is 33.6 million m^3. The basin falls within a warm temperate semi-arid continental monsoon climate zone, with well-defined seasons, ample solar radiation, and moderate rainfall. The annual mean temperature ranges between 12 and 14 °C, with approximately 600 mm of precipitation concentrated in July to September. The You River Reservoir is a medium-scale reservoir that primarily serves urban water supply, flood mitigation and agricultural irrigation. Covering a catchment area of 224 km^2, the reservoir has an average annual runoff of 32.78 million m^3, with a total storage capacity of 30.28 million m^3. The characteristics of upstream, midstream and downstream of the You River are shown in Fig. 3.11.

3.4.2 Shidi River

The Shidi River basin covers an area of 188.68 km^2, with a length of 36.76 km and an average channel slope of 3.62%. The long-term average annual runoff reaches 27.487 million m^3, with a mean flow rate of 0.872 m^3/s. The peak flood discharge during the flood season is 338.3 m^3/s. Two reservoirs, Jijia River Reservoir and Quangou Reservoir, have been constructed on the tributaries. The Shaanxi Fertilizer Plant, situated in the upper reaches of the Shidi River, was constructed in the late 1960s, a period marked by limited environmental technologies, insufficient ecological awareness, and outdated pollution control infrastructure. Consequently, wastewater rich in nitrates and other organic contaminants was discharged directly into the Shidi River, leading to significant degradation of its water quality. In recent years, efforts to control river pollution have involved enhanced environmental regulation of industrial operations. Although the pollution of the Shidi River has shown signs of improvement, legacy contamination remains substantial, and the river is still heavily polluted. The Shidi River is hydraulically connected to the local groundwater system. Extensive groundwater pumping has led to river water infiltration into the aquifer, transporting river-borne pollutants and contributing to groundwater

Fig. 3.12 Shidi River (midstream and downstream from left to right)

contamination, thereby threatening the safety of local drinking water supplies (Wu and Sun 2016). It is estimated that river leakage accounted for 18.4–27.0% of the total groundwater recharge in this area (Li et al. 2016). The characteristics of midstream and downstream of the Shidi River are shown in Fig. 3.12.

3.4.3 Luofu River

The Luofu River originates from Linjiagou and is formed by the convergence of the Daifu, Liu, and Zhu valleys. Flowing northward, it joins the Wei River as a major right-bank tributary. The Luofu River hosts the Luofubao Hydrological Station at its valley mouth, which monitors an upstream watershed area of 122 km^2 along a river reach of 30.5 km. The Luofu River basin is characterized by a warm temperate semi-humid continental climate, with pronounced seasonal variation and alternating wet and dry periods. The long-term mean annual temperature is 13.5 °C. Due to topographic influences, precipitation in the region shows significant spatial variability. Climatic conditions result in uneven intra-annual rainfall distribution, with 53% of annual precipitation concentrated between July and September, leading to marked seasonal variations in river runoff. Rainfall is the primary source of Luofu River recharge, with 70–90% of runoff originating in the southern mountainous region. Consequently, intense rainfall events are the primary driver of flood at the Luofubao Station, with peak flood occurring typically in July and August. In recent years, excessive groundwater extraction has caused significant infiltration of Luofu River water into the groundwater system in the plains, frequently leading to the depletion of flow in the downstream reaches (Fig. 3.13).

Fig. 3.13 Luofu River (upstream, midstream and downstream from left to right)

3.5 Summary

The rivers originating from the northern Qinling Mountains are the lifeblood of the Guanzhong region, providing essential water resources for millions of people and supporting agriculture, industry, and ecosystems. This chapter systematically categorized these rivers based on their geomorphological settings and recharge characteristics, with a focus on those with significant hydrological and socioeconomic roles. In Baoji, rivers like the Qingjiang and Shitou Rivers serve as key drinking water sources. In Xi'an, the Hei and Ba Rivers are vital components of the urban supply network, while the Lao and Feng Rivers illustrate how land use affects water quality and runoff. In Weinan, the You, Shidi, and Luofu Rivers reveal the region's vulnerability to pollution and groundwater overextraction. Case studies and site investigations show that while many rivers maintain good water quality in their upper reaches, pressures from industrialization and urban expansion have degraded downstream sections. Understanding the spatial patterns, runoff characteristics, and ecological conditions of these rivers is critical for developing adaptive, science-based management plans to sustain water availability and quality across the region.

References

Li P, Wu J, Qian H (2016) Preliminary assessment of hydraulic connectivity between river water and shallow groundwater and estimation of their transfer rate during dry season in the Shidi River, China. Environ Earth Sci 75(2):99. https://doi.org/10.1007/s12665-015-4949-7

Li L, Li P, Tian Y, Kou X, He S (2024) Nitrate sources and transformation in surface water and groundwater in Huazhou District, Shaanxi, China: integrated research using hydrochemistry, isotopes and MixSIAR model. Environ Res 263:120052. https://doi.org/10.1016/j.envres.2024.120052

Lu H, Li P, Wang D, Xu F, Fida M (2025) Hydrochemical characteristics and transformation relationship of surface water and groundwater in the Hua County, Guanzhong Plain (China). Water Environ Res 97(6):e70115. https://doi.org/10.1002/wer.70115

Meng X, Xu F (2024) Spatial and temporal characteristics of non-point source pollution and risks in the drinking water source area: a case study of Heihe River in Xi'an. Desalin Water Treat 319:100519. https://doi.org/10.1016/j.dwt.2024.100519

Nsabimana A, Li P, Alam SMK et al (2023) Surface water quality for irrigation and industrial purposes: a comparison between the south and north sides of the Wei River Plain (northwest China). Environ Monit Assess 195:696. https://doi.org/10.1007/s10661-023-11263-0

Ren X, Li P, He X, Zhang Q (2024) Tracing the sources and evaporation fate of surface water and groundwater using stable isotopes of hydrogen and oxygen. Sci Total Environ 931:172708. https://doi.org/10.1016/j.scitotenv.2024.172708

Wu J, Sun Z (2016) Evaluation of shallow groundwater contamination and associated human health risk in an alluvial plain impacted by agricultural and industrial activities, Mid-west China. Expo Health 8:311–329. https://doi.org/10.1007/s12403-015-0170-x

Zhao H, Li P, He X, Ning J (2023) Microplastics pollution and risk assessment in selected surface waters of the Wei River Plain, China. Expo Health 15(4):745–755. https://doi.org/10.1007/s12403-022-00522-z

Zhao W, Qian H, Xu P et al (2024) Tracing groundwater–surface water sources and transformation processes in the Ba River Basin through dual isotopes and water chemistry. Appl Geochem 176:106199. https://doi.org/10.1016/j.apgeochem.2024.106199

Zhi J, Ding A, Zhang S (2016) Nitrate sources and nitrogen biogeochemical processes in the Feng River in West China inferred from the nitrogen and oxygen dual isotope measurements of nitrate. Desalin Water Treat 57:8243–8251. https://doi.org/10.1080/19443994.2015.1021845

Chapter 4
Ecological Baseflow Estimation for Selected Rivers in the Northern Qinling Mountains

Abstract This chapter investigates the estimation of ecological baseflow in small- and medium-sized rivers of the northern Qinling Mountains. Due to increasing anthropogenic pressures and climatic variability, many rivers in the area experience seasonal drying and declining runoff, threatening their ecological integrity. To address this, two primary methods were employed: the hydraulic method, particularly the wetted perimeter method, and the hydrological method using the Tennant approach. The hydraulic analysis involved field surveys at representative river cross-sections to establish wetted perimeter–discharge relationships through power or logarithmic function fitting, identifying ecological baseflow via slope-equals-one and curvature criteria. The hydrological method estimated seasonal baseflow thresholds based on historical runoff data and regional characteristics. The chapter documents the spatial variation of ecological baseflow across river sections and analyzes discrepancies between estimated baseflows and measured discharges. It further explores the ecological status of rivers in Baoji, Xi'an, and Weinan, identifying key drivers of flow interruption, such as uneven precipitation, hydrogeological conditions, and water use patterns. Finally, it proposes targeted strategies for safeguarding ecological baseflow, including monitoring enhancements, adaptive flow management, and policy implementation.

Keywords Baseflow estimation · Wetted perimeter · Historical flow · Ecological status · Ecological flow

4.1 Introduction

Rivers are the cradle of human civilization, sharing an inseparable bond with human society and serving as a vital foundation for its development. However, with population growth and intensified human activities, many rivers worldwide are experiencing phenomena such as unchanged or even increased rainfall but sharply reduced runoff, or even complete drying up. These changes severely impact socio-economic development within river basins and surrounding regions, drawing increasing

P. Li et al., *Water Resources Research in Northern Qinling Mountains, China*, Springer Hydrogeology, https://doi.org/10.1007/978-3-032-09395-0_4

attention to the scientific issue of how human activities, particularly soil and water conservation practices.

River ecological flow is a fundamental element for maintaining river health. With the increasing degradation of ecological and environmental resources, the issue of river ecological flow has gained growing attention. The construction of hydraulic engineering projects, such as sluice gates, dams, reservoirs, and canals, not only alters river morphology and the evolutionary patterns of riverbed geomorphology but also modifies downstream hydrological flow regimes (including discharge magnitude and variability). These changes have triggered ecological crises in rivers, manifested by the weakening of overall ecosystem functions and declines in biodiversity. In recent years, research on river ecological flow has developed rapidly, emerging as one of the most critical global concerns since the twentieth century (Xu et al. 2016). Broadly defined, ecological flow refers to the sufficient water volume and dynamic flow processes that should be retained within river channels to safeguard environmental, social, and economic benefits for downstream regions. In essence, it encompasses the systematic water flow processes required to sustain riverine ecosystem functionality.

Ecological baseflow refers to the minimum discharge required to maintain a river's basic geomorphology, ensure channel conveyance capacity, prevent flow interruption, and preserve a certain level of self-purification capacity (Xu et al. 2016). It represents the critical threshold of perennial in-stream flow necessary to safeguard a river's fundamental ecological functions from degradation. The ecological baseflow is not a fixed value but varies seasonally according to hydrological characteristics, typically differentiated into wet-season and dry-season requirements. Insufficient ecological baseflow often leads to a cascade of eco-environmental issues, including reduced pollution assimilation and dilution capacity, declining groundwater tables, soil erosion, and deterioration of aquatic habitats. Thus, ecological baseflow plays a pivotal role in maintaining riverine ecosystem balance and preserving the integrity and continuity of watershed systems. From a broader perspective, the concept encompasses sustaining the natural morphological structure and evolutionary processes of river channels, supporting the growth, habitat suitability, and reproduction of aquatic organisms, maintaining hydrological connectivity between rivers and associated ecosystems (e.g., floodplains, wetlands), regulating water exchange among atmospheric, surface, and subsurface systems, and preserving self-purification capacity to enable natural pollutant degradation (Xu et al. 2019).

Rivers in northern China exhibit distinct seasonal variations, with flood season runoff accounting for approximately 60–70% of the annual runoff. Therefore, full utilization of flood season water resources should be prioritized while ensuring minimum in-stream flow during dry periods to maintain basic the ecological functions. The northern Qinling Mountains serve as a crucial ecological and water conservation area for the Guanzhong Plain. The abundant water resources represent both a core element and distinctive characteristic of the Qinling ecosystem. This area features diverse geomorphological types and complex geological backgrounds with regional variations, coupled with uneven rainfall distribution. Compounded by unsustainable human activities, these factors have led to environmental quality deterioration, ecosystem degradation, and reduced water resource availability. Long-term data reveal decreasing annual precipitation and runoff in the northern Qinling, where significant pressures on water environment protection exist alongside incomplete ecological compensation mechanisms and numerous adverse ecological impacts from water conservancy projects. Recent ecological surveys indicate that some rivers in the area experience critically low flows or even complete drying up, severely compromising ecosystem health and basic ecological functions. Consequently, conducting ecological baseflow research on small and medium-sized rivers in the northern Qinling Mountains holds substantial practical significance for regional water resource allocation and ecosystem restoration.

The estimation of ecological baseflow in rivers can generally be categorized into four main approaches: hydraulic methods, hydrological methods, eco-environmental simulation methods, and integrated analysis methods. In this study, 55 rivers in the northern Qinling Mountains were investigated and analyzed. Hydraulic and hydrological methods were selected to estimate the ecological baseflow of these rivers.

4.2 Calculation and Analysis of River's Ecological Baseflow

4.2.1 Hydraulic Method-Based Calculation of River's Ecological Baseflow

The hydraulic methods commonly used to calculate river ecological baseflow include the wetted perimeter method and the R2CROSS method. These approaches are grounded in hydraulic theory, and determine ecological flow requirements by analyzing the relationships between aquatic organisms and key hydraulic parameters such as wetted perimeter, channel width, flow velocity, and water depth. River channel hydraulic characteristics are employed as evaluation indicators to assess and establish ecological water demands.

The wetted Perimeter Method is based on the assumption that there is a direct relationship between the wetted perimeter and the suitability of aquatic habitats. It posits that by maintaining the critical wetted perimeter, representing the key habitat area for aquatic organisms, the minimum ecological requirements for non-critical habitat zones can also be generally satisfied. Typically, the wetted perimeter increases with river discharge. However, when the channel morphology transitions from a deep pool to a shallow riffle, a substantial increase in discharge may result in only a marginal increase in wetted perimeter. This inflection point becomes evident on the wetted perimeter–discharge curve. The wetted perimeter method identifies the ecological baseflow as the discharge corresponding to this inflection point on the curve, representing the minimum flow required to sustain in-stream ecological conditions.

The core component of the wetted perimeter method lies in establishing the relationship between wetted perimeter and streamflow. Therefore, the feasibility of constructing a single, stable relationship curve is critical to the success of this method (Ji et al. 2010). Assuming the river is in a state of steady, uniform open-channel flow, the basic uniform flow equation can be derived from the continuity equation and the Chezy formula, as shown in Eqs. (4.1), (4.2) and (4.3):

$$Q = Av \tag{4.1}$$

$$v = C\sqrt{RJ} \tag{4.2}$$

$$Q = AC\sqrt{RJ} \tag{4.3}$$

Where, Q is discharge (m^3/s), A represents the cross-sectional flow area (m^2), v denotes the average velocity of the cross-section (m/s), C is the Chezy coefficient (m), R represents the hydraulic radius (m), and J is hydraulic gradient (dimensionless).

By substituting the Manning formula into the basic uniform flow equation for open-channel flow, the relationship between wetted perimeter and discharge can be derived, as Eqs. (4.4) and (4.5):

$$C = \frac{1}{n} R^{\frac{1}{6}} \tag{4.4}$$

$$Q = \frac{J^{\frac{1}{2}} A^{\frac{5}{3}}}{n P^{\frac{2}{3}}} \tag{4.5}$$

Where, n denotes the roughness coefficient, P represents the wetted perimeter of the river cross-section (in meters), and the other parameters are the same as previously defined.

After establishing a relatively stable wetted perimeter–discharge relationship, the ecological baseflow within the river channel can be estimated based on this curve, with the most critical step being the identification of the change point. From a mathematical perspective, the change point on the curve corresponds to the point with the maximum slope. However, it is generally difficult to locate such a point directly on the wetted perimeter–discharge curve. Gippel and Stewardson (1998) proposed that the change point of the wetted perimeter–discharge curve should occur where the slope equals 1 or at the point of maximum curvature. Furthermore, they indicated that the wetted perimeter–discharge relationship could be fitted using power or logarithmic functions. Triangular, "U"-shaped, and parabolic cross-sections are suited to power function fitting (Eq. 4.6), whereas rectangular and trapezoidal cross-sections are better fitted with logarithmic function (Eq. 4.7).

Power function:

$$y = ax^b + c \tag{4.6}$$

Logarithmic function:

$$y = a\ln x + b \tag{4.7}$$

After obtaining the wetted perimeter and the corresponding discharge data, the location of the change point can be determined. Prior to function fitting, the data need to be normalized to eliminate dimensional units, where $y = P/P_{\max}$ and $x = Q/Q_{\max}$. The coefficients a, b, and c are undetermined parameters to be estimated using the least squares regression method.

Power function and logarithmic function are relatively simple univariate equations, making it easier to determine the critical change point. Typically, the curvature method or slope-equals-one method is used for this purpose. The discharge corresponding to the point of maximum curvature or the point where the slope equals 1 on the curve is considered the ecological flow requirement of the river. The slope equation of the power function is given by $y' = abx^{b-1}$, the slope equation of the logarithmic function is given $y' = a/x$.

The curvature equation corresponding to the power function is given by Eq. (4.8):

$$k = \frac{ab(b-1)x^{b-2}}{\left(1 + a^2b^2x^{2b-2}\right)^{3/2}} \tag{4.8}$$

The curvature equation corresponding to the logarithmic function is given by Eq. (4.9):

$$k = \frac{-ax}{\left(a^2 + x^2\right)^{3/2}} \tag{4.9}$$

When applying the slope method, the point on the wetted perimeter–discharge curve where the slope equals 1 (i.e., where $y' = 1$) is selected. The corresponding discharge at this point is considered the minimum ecological flow within the river channel.

The ecological baseflow equation derived using the slope method based on the power function is given by Eq. (4.10):

$$Q_R = (ab)^{1/(1-b)} \tag{4.10}$$

The ecological baseflow equation derived using the slope method based on the logarithmic function is given by Eq. (4.11):

$$Q_R = a \tag{4.11}$$

The curvature method selects the discharge corresponding to the point of maximum curvature on the curve as the minimum ecological flow within the river channel. To identify this point, the first derivative of the curvature equation must be calculated. The point at which this first derivative equals zero represents the location of maximum curvature. The discharge corresponding to this point is then taken as the minimum ecological flow of the river.

The ecological baseflow equation derived using the maximum curvature method based on the power function is given by Eq. (4.12):

$$Q_R = \left[\frac{2-b}{a^2 b^2 (1-2b)}\right]^{1/2(b-1)} \tag{4.12}$$

The ecological baseflow equation derived using the maximum curvature method based on the logarithmic function is given by Eq. (4.13):

$$Q_R = \frac{a}{\sqrt{2}} \tag{4.13}$$

Based on the calculation principles and formulas of the wetted perimeter method described above, the next step is to organize the measured cross-sectional data to facilitate subsequent calculations. For a given river cross-section, the roughness coefficient n and hydraulic gradient J are considered constants. Therefore, it is only necessary to determine the flow area A corresponding to the wetted perimeter P. The discharge Q associated with each P can then be calculated using the wetted perimeter–discharge relationship.

The roughness coefficient n for the river cross-section can be determined based on empirical values (Huang 2013), as shown in Table 4.1. The hydraulic gradient J at the cross-section can be obtained using regional digital elevation data. By utilizing ArcGIS, two elevation reference points can be identified along the river channel, one upstream and one downstream of the cross-section. The hydraulic gradient J at the cross-section is then calculated based on the elevation difference and the horizontal distance between these two points.

Based on the surveyed geometry of the river cross-section, the flow area corresponding to different wetted perimeters can be calculated. The core of this calculation lies in the application of calculus principles. Once the river cross-sectional curve is obtained, it is discretized for numerical computation, as illustrated in Fig. 4.1. The area enclosed

Table 4.1 Roughness coefficient n for different surface types

Channel bank types	n
Large earth canals under moderate maintenance; small earth canals under good maintenance; small rivers and streams under favorable conditions (free-flowing, without siltation or significant aquatic vegetation)	0.025
Large canals under sub-moderate conditions; small canals under moderate conditions	0.0275
Canals and small rivers under poor conditions (e.g., presence of aquatic plants, gravel beds, partial bank collapse, etc.)	0.030
Canals and small rivers under very poor conditions, with irregular cross-sections and serious obstruction by stones and aquatic vegetation	0.035
Canals and small rivers under extremely poor conditions (e.g., collapsed cliffs, boulders, dense tree roots along the banks, deep pools, landslides, etc.)	0.040

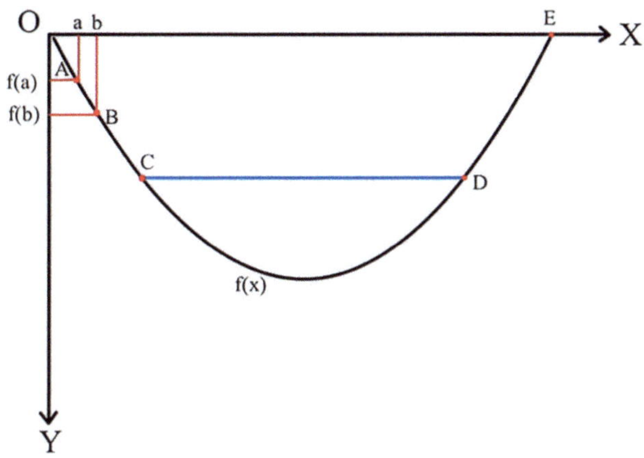

Fig. 4.1 Schematic diagram of wetted perimeter and flow area calculation for a river cross-section

between points A and B on the curve $f(x)$ and the x-axis can be estimated using the trapezoidal rule. The calculation formulae are as follows (Eqs. 4.14 and 4.15):

$$I=\int_a^b f(x)dx \cong \int_a^b f_1(x)dx=\int_a^b\left[f(a)+\frac{f(b)-f(a)}{b-a}(x-a)\right]dx \tag{4.14}$$

$$I=(b-a)\frac{f(a)+f(b)}{2} \tag{4.15}$$

By summing the areas of n small sub-sections, the total area enclosed between the curve $f(x)$ and the x-axis can be obtained, which represents the flow area of the cross-section. At the same time, the distance between points A and B can be calculated using the distance formula between two points in a plane. The total length of the curve $f(x)$, obtained by summing the lengths of all n segments, corresponds to the wetted perimeter of the river. For different wetted perimeters and their corresponding flow areas, the coordinate system can be shifted (for example, moving the origin from point O to point C) to repeat the process and obtain a series of wetted perimeter–area pairs. These values are then substituted into the wetted perimeter–discharge relationship to calculate the corresponding discharge for each wetted perimeter. Based on the resulting dataset, the wetted perimeter–discharge relationship for the cross-section is established, and a fitted curve is plotted accordingly.

By establishing cross-sections along different reaches of the small to medium sized rivers in the northern Qinling Mountains, ecological baseflow calculations were carried out using the wetted perimeter method. The locations of the cross-sections are shown in Fig. 4.2. The cross-section types include parabolic, U-shaped, triangular, and irregular forms. Examples of the cross-sectional shapes and the corresponding curves of relative discharge (Q/Q_{max}) versus relative wetted perimeter (P/P_{max}) are shown in Fig. 4.3 (parabolic cross-section), Fig. 4.4 (U-shaped

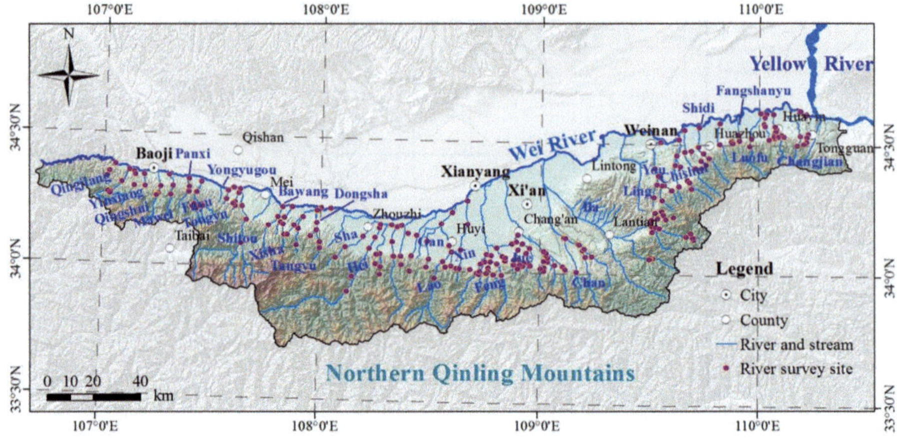

Fig. 4.2 River system and distribution of survey points in the northern Qinling Mountains

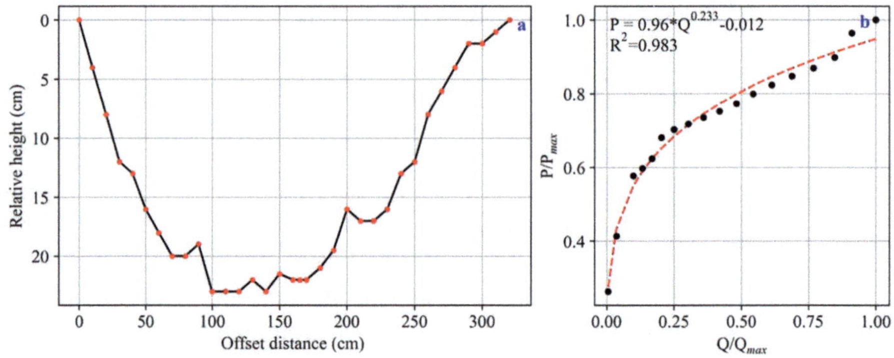

Fig. 4.3 Cross-section shape of N1-QL1 (**a**), and its relative wetted perimeter–relative discharge curve (**b**)

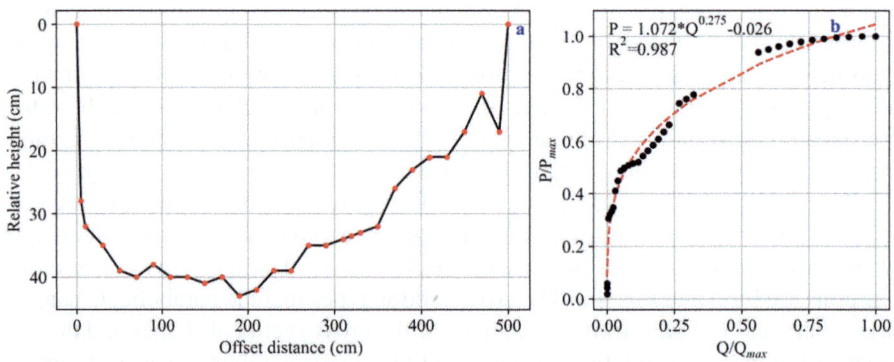

Fig. 4.4 Cross-section shape of N2-QL13 (**a**), and its relative wetted perimeter–discharge curve (**b**)

cross-section), Fig. 4.5 (triangular cross-section), Fig. 4.6 (approximately trapezoidal cross-section), and Fig. 4.7 (irregular cross-section). Detailed calculation results are provided in Tables 4.2, 4.3, and 4.4.

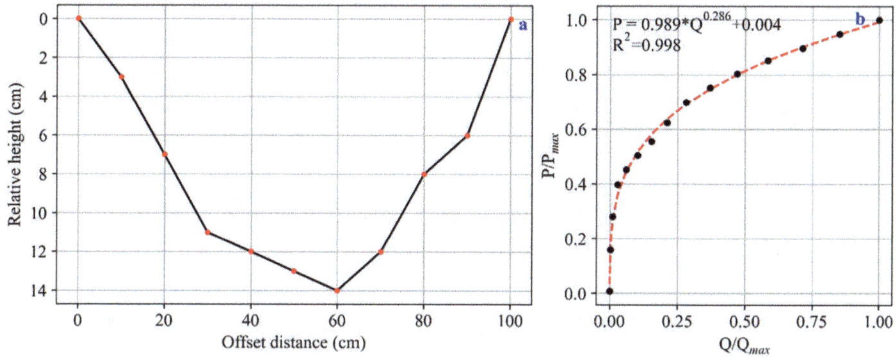

Fig. 4.5 Cross-section shape of N1-QL11 (**a**), and its relative wetted perimeter–relative discharge curve (**b**)

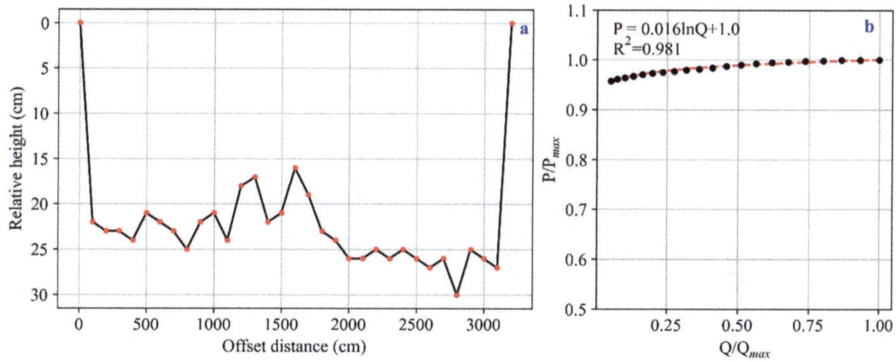

Fig. 4.6 Cross-section shape of N2-QL72 (**a**) and its relative wetted perimeter–relative discharge curve (**b**)

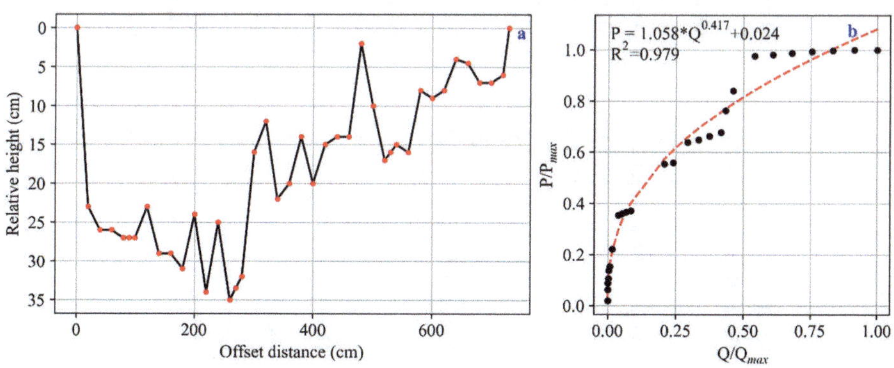

Fig. 4.7 Cross-section shape of N1-QL4 (**a**) and its relative wetted perimeter–relative discharge curve (**b**)

Table 4.2 Calculation of ecological baseflow for small- and medium-sized rivers in the Baoji region

Rivers	ID	Location	Cross-sectional shape	Wetted perimeter–discharge curve fitting		Ecological baseflow (m^3/s)	
				Equations	R^2	Slope-equals-one	Curvature method
Taiyin River	N1-QL1	Downstream	Parabola	$P = 0.96Q^{0.233} - 0.012$	0.9830	0.0373	0.0171
	N1-QL2	Midstream	Parabola	$P = 1.051Q^{0.247} - 0.012$	0.9910	0.0496	0.0218
	N1-QL3	Upstream	Parabola	$P = 0.971Q^{0.413} + 0.038$	0.9950	0.0191	0.0029
Qingjiang River	N1-QL4	Downstream	Irregular shape	$P = 1.058Q^{0.417} + 0.024$	0.9790	0.0914	0.0133
	N1-QL5	Midstream	Parabola	$P = 0.878Q^{0.363} + 0.071$	0.9876	0.1668	0.0411
	N1-QL6	Upstream (after confluence)	Irregular shape	$P = 1.049Q^{0.246} - 0.002$	0.9761	0.1484	0.0653
	N1-QL18	Upper tributary 1	Approximately U-shape	$P = 1.191Q^{0.239} - 0.165$	0.9964	0.0439	0.0197
	N1-QL19	Upper tributary 2	Irregular shape	$P = 1.041Q^{0.43} + 0.007$	0.9884	0.5704	0.0689
Yinxiang River	N1-QL7	Lower section of minor tributary	Approximately triangle	$P = 0.941Q^{0.668} + 0.074$	0.9761	0.0176	/
	N1-QL8	Downstream	Approximately U-shape	$P = 0.582Q^{0.662} + 0.465$	0.9641	0.0112	/
	N1-QL9	Upstream	Approximately U-shape	$P = 0.776Q^{0.848} + 0.265$	0.9936	0.0914	/
Dongcha River	N1-QL10	Downstream	Approximately triangle	$P = 0.882Q^{0.894} + 0.143$	0.9560	0.0356	/
	N1-QL11	Upstream	Triangle	$P = 0.989Q^{0.286} + 0.004$	0.9976	0.0246	0.0093
	N1-QL12	Midstream	Irregular shape	$P = 1.135Q^{0.33} - 0.034$	0.9779	0.2063	0.0628
Qingshui River	N1-QL14	After upstream confluence	Irregular shape	$P = 1.119Q^{0.567} + 0.006$	0.9505	0.2964	/
	N1-QL15	Upstream tributary 1	Approximately triangle	$P = 1.14Q^{0.258} - 0.096$	0.9823	0.3162	0.1336
	N1-QL16	Upstream tributary 2	Irregular shape	$P = 1.228Q^{0.221} - 0.162$	0.9882	0.2791	0.1327

Mawei River	N1-QL13	Downstream	Irregular shape	$P = 1.298Q^{0.225} - 0.238$	0.9505	0.0931	0.0437
	N1-QL21	Midstream	Approximately trapezium	$P = 0.799Q^{0.651} + 0.254$	0.9629	0.0458	/
	N1-QL20	Upstream (at the mountain outlet)	Approximately triangle	$P = 1.148Q^{0.399} - 0.073$	0.9906	0.0277	0.0050
	N2-QL17	Midstream	Approximately triangle	$P = 1.133Q^{0.406} - 0.043$	0.9465	0.0793	0.0132
	N2-QL16	Upstream	Approximately U-shape	$P = 1.305Q^{0.201} - 0.226$	0.9379	0.5154	0.2559
Fayu River	N2-QL8	Downstream	Irregular shape	$P = 1.045Q^{0.233} - 0.04$	0.9833	0.0433	0.0198
	N2-QL9	Midstream	Triangle	$P = 1.031Q^{0.323} - 0.013$	0.9926	0.2134	0.0675
	N2-QL12	Upstream first-order tributary 1	Triangle	$P = 1.035Q^{0.389} - 0.002$	0.9934	0.1412	0.2780
	N2-QL10	Upstream second-order tributary 1	Irregular shape	$P = 1.14Q^{0.481} - 0.021$	0.9546	0.1159	0.0034
	N2-QL11	Upstream second-order tributary 2	Approximately U-shape	$P = 1.337Q^{0.153} - 0.24$	0.9441	0.1309	0.0735
	N2-QL13	Upstream first-order tributary 2	U-shape	$P = 1.0272Q^{0.275} - 0.026$	0.9866	0.4014	0.1590
	N2-QL14	Upstream second-order tributary 3	Triangle	$P = 0.94Q^{0.424} + 0.062$	0.9923	0.0767	0.0101
	N2-QL15	Upstream second-order tributary 4	Approximately parabola	$P = 1.041Q^{0.255} - 0.059$	0.9891	0.1708	0.0728
Tongyu River	N2-QL6	Downstream	Triangle	$P = Q^{0.375}$	0.9999	0.0931	0.0208
	N2-QL7	Midstream	Triangle	$P = Q^{0.375}$	0.9999	0.0555	0.0124
Maili River	N2-QL2	Downstream	Triangle	$P = 0.953Q^{0.355} + 0.01$	0.9982	0.0390	0.0102
	N2-QL3	Upstream	Approximately U-shape	$P = 1.613Q^{0.106} - 0.61$	0.9997	0.0104	0.0064
	N2-QL5	Upstream tributary 1	Approximately triangle	$P = 1.122Q^{0.309} - 0.052$	0.9822	0.0183	0.0063
	N2-QL4	Upstream tributary 2	Approximately parabola	$P = 1.095Q^{0.24} - 0.039$	0.9778	0.0264	0.0119

Note: "/" indicates that calculation is not possible ($b > 0.5$)

Table 4.3 Calculation of ecological baseflow for small- and medium-sized rivers in the Xi'an region

Rivers	ID	Location	Cross-sectional shape	Wetted perimeter–discharge curve fitting		Ecological baseflow (m^3/s)	
				Equations	R^2	Slope-equals-one	Curvature method
Zigeyu	N1-QL67	Upstream	Irregular shape	$P = 1.084Q^{0.372} - 0.019$	0.9906	0.1360	0.0311
	N1-QL65	Upstream	Irregular shape	$P = 1.089Q^{0.288} - 0.087$	0.9806	0.3387	0.1270
Shenshuiyu	N1-QL66	Upstream	Irregular shape	$P = 1.204Q^{0.204} - 0.179$	0.9949	0.0835	0.0416
Gaoguanyu	N2-QL69	Mountain exit	Irregular shape	$P = 0.137\ln Q + 0.984$	0.9734	0.1604	0.1134
	N2-QL72	Upstream	Approximately trapezium	$P = 0.016\ln Q + 1.0$	0.9814	0.0226	0.0160
Xiangyu	N2-QL70	Downstream	Irregular shape	$P = 1.372Q^{0.176} - 0.302$	0.9619	0.3906	0.2083
	N2-QL71	Upstream	Approximately U-shape	$P = 1.139Q^{0.164} - 0.049$	0.9091	0.2058	0.1129
Jinsha River	N2-QL83	Downstream	Approximately U-shape	$P = 1.175Q^{0.192} - 0.137$	0.9945	0.3089	0.1587
	N2-QL80	Upstream	Approximately U-shape	$P = 1.077Q^{0.241} - 0.086$	0.9778	0.0866	0.0387
Ziwuyu River	N2-QL84	Mountain exit	Approximately U-shape	$P = 1.015Q^{0.213} - 0.077$	0.9806	0.5167	0.2514
Baolongyu River	N2-QL76	Midstream	Approximately U-shape	$P = 1.456Q^{0.131} - 0.414$	0.9509	0.8628	0.5053
	N2-QL75	Downstream	Approximately triangle	$P = 1.16Q^{0.32} - 0.085$	0.9563	0.5436	0.1747
Tianziyu River	N1-QL80	Upstream	Approximately U-shape	$P = 1.1Q^{0.303} - 0.084$	0.9907	0.2958	0.1039
	N1-QL79	Midstream	Irregular shape	$P = 1.149Q^{0.307} - 0.057$	0.9732	0.1471	0.0507
Hao River	N1-QL81	Upstream	Irregular shape	$P = 1.256Q^{0.213} - 0.194$	0.9792	0.1651	0.0802
	N1-QL83	Midstream	Approximately parabola	$P = 2.521Q^{0.055} - 1.497$	0.9884	0.1328	0.0878
	N2-QL90	Downstream	Approximately triangle	$P = 2.892Q^{0.056} - 1.902$	0.9819	0.2938	0.1941
Jue River	N2-QL85	Midstream	Irregular shape	$P = 1.239Q^{0.19} - 0.183$	0.9911	0.4294	0.2216
Tianyiyu	N2-QL74	Upstream	Approximately U-shape	$P = 1.389Q^{0.184} - 0.314$	0.9734	4.6795	2.4498
	N1-QL71	Midstream	Approximately U-shape	$P = 1.197Q^{0.161} - 0.175$	0.9953	0.9116	0.5029
Jiaoyu River	N1-QL68	Midstream	Irregular shape	$P = 1.145Q^{0.33} - 0.071$	0.9905	0.2268	0.0691
Xiaoyu River	N1-QL75	Downstream	Approximately U-shape	$P = 1.263Q^{0.175} - 0.228$	0.9837	0.5744	0.3071
	N1-QL72	Upstream	Irregular shape	$P = 1.191Q^{0.212} - 0.171$	0.9941	0.2865	0.1394

Dayu River	N1-QL74	Upstream	Approximately U-shape	$P = 1.003Q^{0.319} - 0.014$	0.9954	0.8172	0.2640
	N1-QL73	Upstream	Irregular shape	$P = 1.213Q^{0.183} - 0.188$	0.9918	0.4825	0.2533
Kuyu River	N2-QL91	Upstream	Approximately U-shape	$P = 1.217Q^{0.217} - 0.157$	0.9848	1.6065	0.7723
Tangyu River	N2-QL94	Downstream	Approximately U-shape	$P = 1.455Q^{0.106} - 0.46$	0.9878	0.4016	0.2460
Chan River	N1-QL89	Upstream	Approximately parabola	$P = 1.237Q^{0.164} - 0.207$	0.9931	0.2731	0.1496
	N2-QL96	Upstream	Approximately parabola	$P = 1.109Q^{0.123} - 0.106$	0.9945	0.6628	0.3942
	N2-QL95	Upstream	Approximately triangle	$P = 1.059Q^{0.238} - 0.049$	0.9938	0.4013	0.1810
	N2-QL92	Upstream	Approximately triangle	$P = 1.077Q^{0.411} + 0.002$	0.986	0.3405	0.0528
Ba River	N2-QL64	Upstream tributary	Approximately U-shape	$P = 0.954Q^{0.435} + 0.044$	0.9709	0.2068	0.0228
	N2-QL65	Upstream	Approximately triangle	$P = 1.134Q^{0.387} - 0.02$	0.9573	0.5987	0.1200
	N2-QL66	Upstream tributary	Approximately triangle	$P = 1.261Q^{0.17} - 0.275$	0.9859	0.1158	0.0627
	N1-QL56	Upstream	Irregular shape	$P = 1.159Q^{0.386} - 0.042$	0.9608	0.4172	0.0846
	N1-QL57	Upstream	Approximately U-shape	$P = 1.103Q^{0.223} - 0.094$	0.9877	0.1544	0.0730
	N1-QL55	Upstream	Approximately U-shape	$P = 1.275Q^{0.165} - 0.2$	0.9573	0.1925	0.1052
	N1-QL52	Upstream	Irregular shape	$P = 1.016Q^{0.337} + 0.025$	0.9812	0.0653	0.0191
	N1-QL53	Upstream	Irregular shape	$P = 1.15Q^{0.434} - 0.021$	0.9647	0.1861	0.0209
Qingyugou River	N2-QL59	Downstream	Approximately U-shape	$P = 1.311Q^{0.141} - 0.276$	0.9826	0.0987	0.0568
	N2-QL58	Downstream	Irregular shape	$P = 1.192Q^{0.302} - 0.109$	0.9862	0.1304	0.0460
	N2-QL55	Tributary	Approximately U-shape	$P = 3.064Q^{0.032} - 2.063$	0.9998	0.0312	0.0213
	N2-QL56	Midstream	Approximately triangle	$P = 1.146Q^{0.329} - 0.058$	0.9779	0.1239	0.0380
	N2-QL57	Upstream	Approximately U-shape	$P = 1.06Q^{0.32} - 0.032$	0.9870	0.0883	0.0285
Tongyu River	N2-QL62	Downstream	Irregular shape	$P = 0.217\ln Q + 1.004$	0.9799	0.0608	0.0430
	N2-QL60	Midstream	Irregular shape	$P = 1.157Q^{0.223} - 0.167$	0.9915	0.0576	0.0272
	N2-QL61	Upstream	Approximately triangle	$P = 1.045Q^{0.356} - 0.016$	0.9905	0.1117	0.0289

Table 4.4 Calculation of ecological baseflow for small- and medium-sized rivers in the Weinan region

Rivers	ID	Location	Cross-sectional shape	Wetted perimeter–discharge curve fitting		Ecological baseflow (m^3/s)	
				Equations	R^2	Slope-equals-one	Curvature method
You River	N2-QL37	Midstream	Approximately triangle	$P = 1.14Q^{0.267} - 0.102$	0.9932	0.1968	0.0803
	N2-QL34	Tributary downstream	Approximately U-shape	$P = 1.29Q^{0.099} - 0.232$	0.9188	0.0381	0.0236
	N2-QL35	Tributary midstream	Irregular shape	$P = 1.157Q^{0.459} - 0.008$	0.9273	0.2764	0.0182
	N1-QL37	Midstream	Irregular shape	$P = 1.241Q^{0.43} - 0.065$	0.9456	0.1698	0.0205
	N1-QL38	Midstream	Approximately triangle	$P = 1.166Q^{0.382} - 0.056$	0.9568	0.3235	0.0682
	N1-QL39	Upstream	Approximately parabola	$P = 1.146Q^{0.336} - 0.064$	0.9864	0.2073	0.0609
	N1-QL36	Tributary downstream	Irregular shape	$P = 1.268Q^{0.226} - 0.176$	0.9543	0.0989	0.0463
Chishui River	N2-QL27	Tributary downstream	Approximately U-shape	$P = 1.089Q^{0.219} - 0.076$	0.9708	0.0761	0.0364
	N2-QL26	Tributary upstream	Approximately triangle	$P = 1.024Q^{0.452} + 0.019$	0.9883	0.1663	0.0131
	N2-QL28	Midstream	Approximately U-shape	$P = 1.144Q^{0.218} - 0.117$	0.9655	0.3458	0.1655
	N2-QL29	Upstream	Approximately parabola	$P = 1.022Q^{0.376} + 0.028$	0.9915	0.1630	0.0361
	N2-QL30	Tributary upstream	Approximately parabola	$P = 1.164Q^{0.24} - 0.11$	0.9808	0.3315	0.1487
	N2-QL31	Upstream	Approximately triangle	$P = 1.006Q^{0.329} - 0.043$	0.9897	0.1645	0.0506
Yuxian River	N1-QL24	Downstream	Approximately parabola	$P = 1.091Q^{0.228} - 0.084$	0.9757	0.0175	0.0081
	N1-QL34	Upstream	Irregular shape	$P = 1.141Q^{0.513} + 0.005$	0.9474	0.0894	/
	N1-QL27	Tributary downstream	Irregular shape	$P = 1.299Q^{0.273} - 0.177$	0.9155	0.0460	0.0184

Shidi River	N1-QL23	Downstream	Approximately parabola	$P = 0.951Q^{0.371} + 0.02$	0.9902	0.0278	0.0064
	N1-QL29	Tributary	Approximately U-shape	$P = 0.982Q^{0.271} + 0.006$	0.9929	0.0400	0.0161
	N1-QL30	Tributary upstream	Approximately U-shape	$P = 1.496Q^{0.204} - 0.484$	0.9914	0.0596	0.0297
	N1-QL32	Upstream	Approximately parabola	$P = 1.175Q^{0.219} - 0.142$	0.9889	0.1843	0.0882
Gouyu River	N2-QL22	Mountain exit	Approximately U-shape	$P = 1.214Q^{0.204} - 0.183$	0.9712	0.0264	0.0131
	N2-QL23	Upstream	Approximately parabola	$P = 1.308Q^{0.075} - 0.3$	0.9988	0.0215	0.0138
Fangshan River	N2-QL20	Midstream	Triangle	$P = 1.024Q^{0.312} - 0.013$	0.9955	0.0162	0.0054
	N2-QL18	Mountain exit	Triangle	$P = 0.861Q^{0.528} + 0.101$	0.9979	0.0832	/
	N2-QL19	Upstream	Approximately triangle	$P = 1.084Q^{0.443} - 0.026$	0.9883	0.1166	0.0111
Luofu River	N1-QL40	Downstream	Approximately U-shape	$P = 1.103Q^{0.167} - 0.142$	0.916	0.0121	0.0066
	N1-QL41	Midstream	Approximately triangle	$P = 0.852Q^{0.572} + 0.088$	0.9775	0.0904	/
	N1-QL42	Upstream	Irregular shape	$P = 1.136Q^{0.394} - 0.035$	0.9476	0.4587	0.0865
Liuye River	N2-QL51	Midstream	Approximately triangle	$P = 1.085Q^{0.447} - 0.033$	0.981	0.0041	0.0004
Changjian River	N2-QL47	Downstream	Approximately U-shape	$P = 1.079Q^{0.268} - 0.061$	0.993	0.0346	0.0141
Baimajian River	N2-QL44	Upstream	Approximately parabola	$P = 1.281Q^{0.162} - 0.277$	0.9834	0.2722	0.1500
Shuiyu River	N2-QL38	Downstream	Approximately U-shape	$P = 1.584Q^{0.087} - 0.643$	0.9896	0.0048	0.0031
	N2-QL42	Upstream	Triangle	$P = Q^{0.375}$	0.9999	0.0216	0.0048

Note: "/" indicates that calculation is not possible ($b > 0.5$)

By comparing the ecological baseflow values calculated using the slope-equals-one method and the curvature method, it was found that the two methods yield significantly different estimates. Generally, the ecological baseflow estimated by the slope-equals-one method is higher than that obtained by the curvature method, consistent with previous research findings (Liu et al. 2006; Ji et al. 2006). When fitting with a logarithmic function, the ratio of ecological baseflow determined by the slope-equals-one method to that determined by the curvature method is constant ($\sqrt{2}$). In contrast, when fitting with a power function, this ratio varies as a constant depending on the exponent b (Ji et al. 2010). Under both fitting relationships, the ecological baseflow estimated by the slope-equals-one method is larger than that estimated by the curvature method, and this difference correlates with both the shape and size of the cross-section (Ji et al. 2006).

Currently, there is no universally accepted criterion for selecting between the slope-equals-one method and the curvature method (Ji 2006; Ji et al. 2010). Based on practical calculations, it was observed that when fitting with a power function, some fitted exponents b exceed 0.5, rendering the curvature method unsuitable for calculating ecological baseflow. Moreover, comparisons between curvature method estimates and measured flows indicate that, for some cross-sections, ecological baseflow values estimated by the curvature method are significantly lower than the observed flows. If the curvature method values were used as ecological baseflow, river depths would be unrealistically shallow, which contradicts actual conditions. Therefore, this study selects the ecological baseflow estimated by the slope-equals-one method as the representative ecological baseflow for the rivers.

The ecological baseflow calculation results also show that the estimated ecological baseflow varies between upstream and downstream sections of the same river, and the differences also vary across different rivers. In some rivers, the ecological baseflow estimated at upstream cross-sections is greater than that at downstream cross-sections, while in other cases the opposite is observed. For upstream cross-sections, the hydraulic gradient is generally greater than that of downstream sections; therefore, if the cross-sectional shapes of the upstream and downstream sections are roughly the same, the ecological baseflow in the upstream section is expected to be greater than that in the downstream. However, from a practical perspective, natural rivers tend to have narrower upstream sections and wider downstream sections, and since the shape and size of the cross-section influence the wetted perimeter–discharge relationship, such variation leads to differences in the estimated ecological baseflow between upstream and downstream sections of the same river. This also indirectly indicates that, in order to ensure ecological baseflow requirements are met, different sections of the same river should be assigned different threshold values.

4.2.2 Hydrological Method-Based Calculation of River Ecological Baseflow

The hydrological method, also known as the historical flow method, is based on historical hydrological data of rivers, and determines ecological flow using simple hydrological indicators. This method is easy to apply and does not require field measurements; however, it does not consider ecological habitats, water quality, water temperature, seasonal variation, landscape features, or changes in riverbed morphology. Hydrological methods include the Tennant method, the 7Q10 method, and the Texas method, among which the Tennant method is the most widely used, though it has relatively low accuracy. The application of this method relies on a good understanding of the region's ecological and topographical characteristics. It is necessary to assess whether the recommended flow proportions are suitable for local river conditions and to adjust the proportions based on the local river management objectives. Therefore, the recommended proportion may vary across regions and seasons. A major limitation of this method is that it oversimplifies the actual river conditions and does not take into account the ecological needs of the aquatic ecosystem. As such, it is only suitable for river reaches with low ecological priority or for use as a preliminary screening tool in combination with other methods.

Xu et al. (2019) conducted a comprehensive comparison of seven hydrological methods for calculating baseflow in the Wei River Basin and found that the baseflow estimated using the Tennant method was generally consistent with the minimum ecological flow specified in the Recent Key Management Plan for the Wei River Basin of that year. Based on this consistency, the Tennant method was considered relatively reasonable. The small- to medium-sized rivers investigated in this study, located in the northern Qinling Mountains, are all tributaries of the Wei River, and the regional environmental conditions are relatively similar. Therefore, the Tennant method was adopted in this section to estimate the ecological baseflow of these rivers. The Tennant method, proposed by Tennant in 1976, was based on detailed field studies conducted between 1964 and 1974 by Tennant and colleagues on 11 rivers in the United States. The purpose was to test the effectiveness of the Tennant method by examining the physical, chemical, and biological characteristics of rivers in different regions, with different channel types, under varying flow conditions, and to assess their impacts on both coldwater and warmwater fisheries. Their study concluded that:

1. Ten percent of the average annual flow is recommended as the minimum instantaneous flow to support short-term survival habitat for most aquatic organisms. At this flow level, channel width, water depth, and velocity are significantly reduced, aquatic habitats become degraded, and nearly half of the streambed or wetted perimeter is exposed. Tributary channels may be severely dewatered or completely dry. To maintain essential ecological functions such as fish habitat, spawning, and nursery areas, a minimum level of surface flow in the river channel must be ensured in order to provide suitable shallow water areas and sufficient depth.

2. For typical rivers, when the in-channel flow is between 60 and 100% of the average annual flow, the channel width, depth, and velocity provide excellent conditions for the growth of aquatic organisms. Most riffles and shallow areas are submerged, with only a few cobbles or sandbars exposed. The banks and adjacent floodplains become accessible to fish, and riparian vegetation receives adequate water. Invertebrate species are diverse and abundant.
3. When the in-channel flow accounts for 30–60% of the average annual flow, the channel width, depth, and velocity are generally satisfactory. Most shallow areas, except for extremely wide gravel bars, are submerged, and most side channels contain flowing water. Many riverbank zones can serve as active areas for fish.
4. In the case of large rivers, even when the flow is only 5–10% of the average annual flow, the channel still maintains a certain width, depth, and velocity, which can generally meet the basic requirements for fish migration, survival, recreation, and landscape maintenance. This flow is considered the minimum instantaneous discharge necessary to support the short-term survival of the vast majority of aquatic organisms.

The Tennant method considers 10% of the multi-year average flow as the minimum threshold for maintaining in-channel flow, which helps prevent river disconnection and avoids irreversible damage to aquatic communities. Ecological baseflow is closely related to the evolution of river ecosystems and the life history of aquatic organisms. Since the growth of aquatic organisms is influenced by factors such as water quantity and temperature, ecological baseflow is not a fixed value throughout the year. According to the hydrological characteristics of rivers, ecological baseflow is generally divided into flood season and non-flood season values. Under typical conditions, the ecological baseflow during the non-flood season should not be lower than 10% of the multi-year average natural runoff, while during the flood season, it is recommended to reach 20–30% of the multi-year average natural runoff. Therefore, to ensure more reasonable ecological baseflow values, this section adopts 30% of the multi-year average flow as the ecological baseflow from April to September, and 10% of the multi-year average flow as the ecological baseflow from October to March of the following year.

Since hydrological methods require historical hydrological data of rivers, it is first necessary to select rivers with established hydrological stations, which are typically medium or large rivers. Based on publicly available information from government departments, this study organized and analyzed the *Water Resources Bulletins* of Shaanxi Province (2011–2019), Baoji City (2011–2019), and Xi'an City (2010–2019) (hydrological data for Weinan City have not yet been collected). As a result, the multi-year average runoff of four rivers in the Baoji section of the northern Qinling Mountains (Qingjiang River, Fayu River, Shitou River, and Tangyu River) was obtained.

The multi-year average runoff data in this section are based on the 2019 data published in the *Baoji Water Resources Bulletin*. The multi-year average runoff values of four typical small- and medium-sized rivers in Baoji City are shown in Table 4.5.

Table 4.5 Multi-year average runoff of small- and medium-sized rivers in the Baoji section of the northern Qinling Mountains (10^8 m^3)

River in Baoji section	Qingjiang River	Fayu River	Shitou River	Tangyu River
Multi-year average runoff	1.989	3.611	4.082	3.500

Table 4.6 Ecological baseflow calculation for small- and medium-sized rivers in the Baoji section of the northern Qinling Mountains (10^8 m^3)

Month	Qingjiang River	Fayu River	Shitou River	Tangyu River
April to September (30%)	0.5967	1.0833	1.2246	1.05
October to March (10%)	0.1989	0.3611	0.4082	0.35

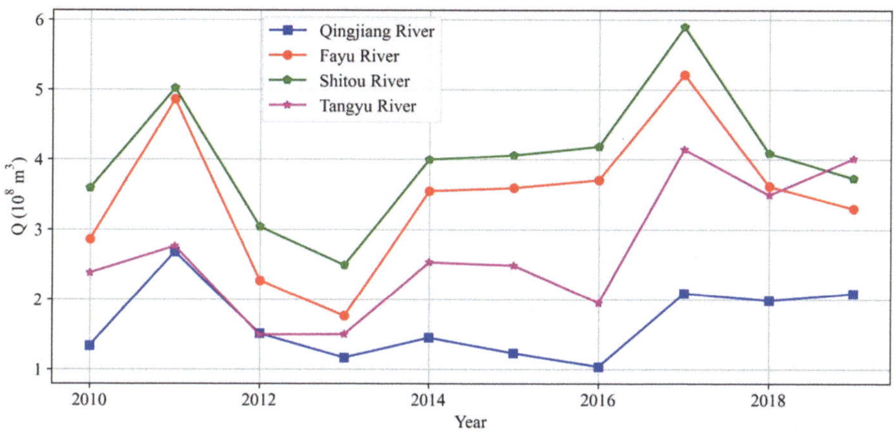

Fig. 4.8 Annual runoff variation of typical rivers in the Baoji section over the past decade

The ecological baseflow during flood and non-flood seasons, calculated based on the multi-year average runoff data of four typical small- and medium-sized rivers in Baoji City, is shown in Table 4.6.

Figure 4.8 shows the annual runoff variations over the past decade (2010–2019) for four typical rivers in the Baoji region: Qingjiang River, Fayu River, Shitou River, and Tangyu River. From the runoff trend graphs of these rivers, it can be seen that all four rivers exhibit similar variation patterns, with peak flow values occurring around 2011 and 2017. This is likely due to increased precipitation during those years, which led to a general increase in river flows across the region. Minor differences may be attributed to localized human activities. The river flows sharply decreased in 2012 and 2013, followed by a rising trend starting in 2013, although the flows did not reach the 2011 levels. Starting from 2016, flows increased again, reaching their highest point in 2017, and then remained relatively stable in the subsequent 2 years. The similar trends among these four rivers indicate that river flows are primarily controlled by local precipitation, resulting in consistent runoff variation patterns across rivers in the same area. This suggests a clear relationship between regional river runoff and local rainfall. Understanding the annual runoff variations of regional rivers allows for analysis of the underlying causes, which can support simple forecasting of future river flows and provide a basis for measures to address flow reductions in certain rivers.

4.2.3 Analysis of Ecological Baseflow Results

Ecological baseflow is the minimum flow required to sustain key species necessary for maintaining ecosystem balance. Prolonged or frequent river flows below the ecological baseflow can lead to degradation of river ecological functions and reduced self-purification capacity. In this study, the ecological baseflow of rivers was estimated using the wetted perimeter method. Previous research has shown that when the wetted perimeter–discharge relationship follows a power function, the slope-equals-one method is more appropriate for estimation (Zhang et al. 2021). Therefore, after fitting the curve of relative wetted perimeter versus relative discharge, the ecological baseflow estimated by the slope-equals-one method was compared with measured flow data.

In the northern Qinling Mountains, ten rivers in the Baoji section (Taiyinyu River, Qingjiang River, Yinxiang River, Dongcha River, Qingshui River, Mawei River, Panxi River, Fayu River, Tongyu River, and Maili River) were surveyed, and several representative cross sections were measured. The distribution of measured discharge and estimated ecological baseflow at different cross sections is shown in Fig. 4.9. Flow measurements at river sections were conducted from August 1 to August 4, during which no rainfall occurred before or during the measurements. As shown in Fig. 4.9, most sections basically meet the ecological baseflow estimated

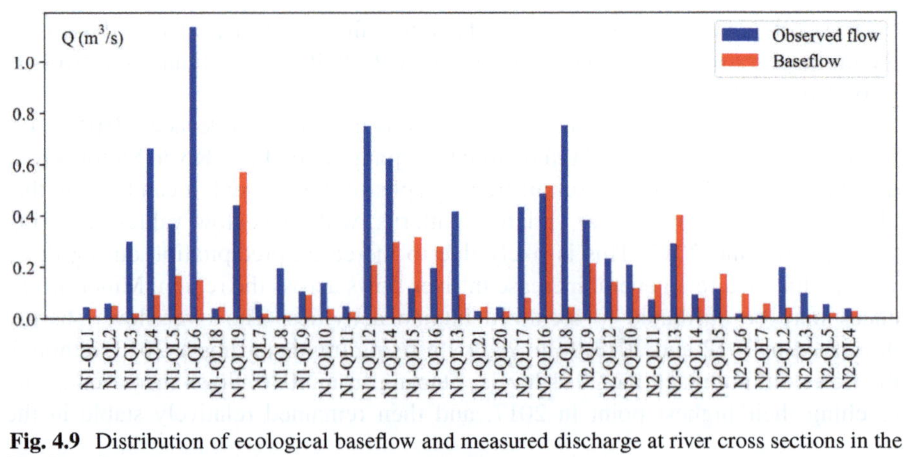

Fig. 4.9 Distribution of ecological baseflow and measured discharge at river cross sections in the Baoji section

using the wetted perimeter method. In some rivers (Taiyinyu River, Yinxiang River, Dongcha River, Qingshui River, Fayu River, and Maili River), measured discharge at certain sections is approximately equal to the estimated ecological baseflow. No flow interruption was found in any of the surveyed rivers in the Baoji section. However, some sections, such as the upstream tributaries of the Qingjiang River, Qingshui River, Fayu River, and Tongyu River, had measured discharge values lower than the estimated ecological baseflow. Notably, none of the selected sections in the Tongyu River met the estimated ecological baseflow requirements.

The 22 rivers surveyed in the Xi'an section of the northern Qinling Mountains include Taipingyu River, Zigeyu River, Shenshuiyu River, Gaoguanyu River, Xiangyu River, Feng River, Jinsha River, Ziwuyu River, Baolongyu River, Tianziyu River, Hao River, Yu River, Taiyiyu River, Jiaoyu River, Xiaoyu River, Dayu River, Kuyuyu River, Tangyu River, Chan River, Ba River, Qingyugou River, and Tongyu River. The flow measurements at river cross sections were conducted between August 10 and August 17. A significant rainfall event during the measurement period led to a marked increase in river discharge. In addition, due to safety concerns stemming from mountain closures at some valleys, measurements were not conducted in certain rivers with high flows following the rainfall. Field investigations revealed that, among the small- to medium-sized rivers in the Xi'an, only the downstream of the Shibianyu Reservoir experienced flow interruption. According to the measured discharge data and the estimated ecological baseflow values (Fig. 4.10), most river cross sections exhibited measured flows that met or exceeded the ecological baseflow estimated using the wetted perimeter method. However, in some rivers, specifically Ziwuyu River, Baolongyu River, Tianziyu River, and Dayu River, certain cross sections had measured flows that fell short of the estimated ecological baseflow. A comparison showed that most of the cross sections with insufficient flows were located in the upstream segments of the rivers, with a few in the middle reaches.

The field investigation in the Weinan section of the northern Qinling Mountains covered 11 rivers (Qiu River, Chishui River, Yuxian River, Shidi River, Gouyu

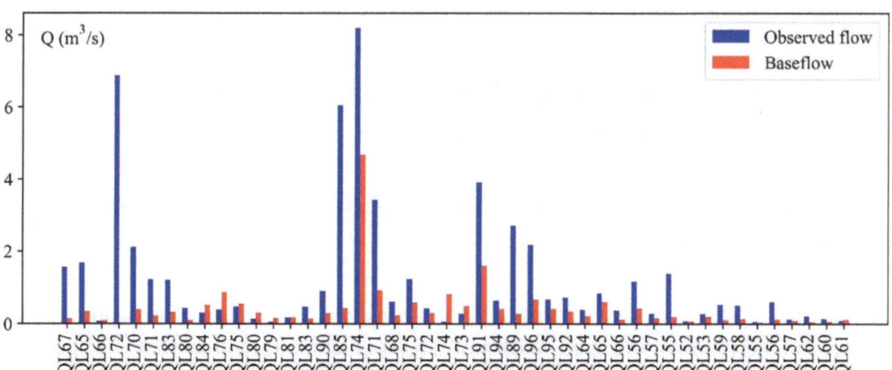

Fig. 4.10 Distribution of ecological baseflow and measured discharge at river cross sections in the Xi'an section

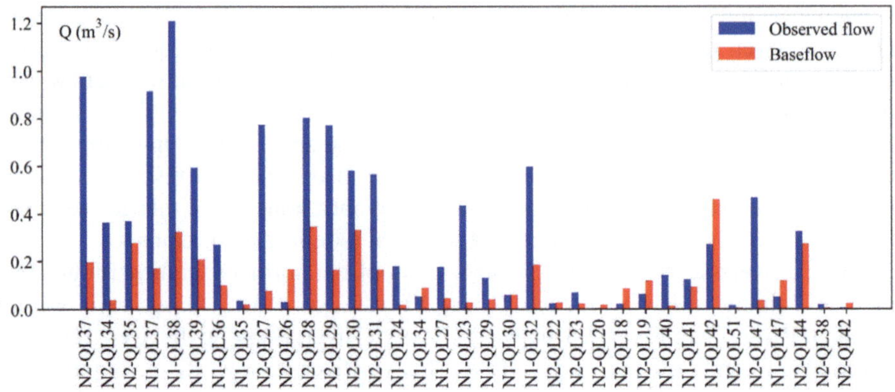

Fig. 4.11 Distribution of ecological baseflow and measured discharge at each cross section of rivers in the Weinan section

River, Fangshan River, Luofu River, Liuye River, Changjian River, Baimajian River, and Shuiyu River), conducted from August 4 to 9. No significant rainfall occurred prior to or during the investigation. Field observations revealed that nearly one-third of the surveyed cross sections in this section were dry, indicating interrupted flow. The Liuye, Changjian, Baimajian, and Shuiyu rivers were found to be almost entirely dry along their lengths, while the Yuxian, Shidi, and Gouyu rivers exhibited dry conditions at some cross sections. Based on the comparison between measured discharges and the estimated ecological baseflows (Fig. 4.11), certain upstream sections of the Yuxian River, upstream and midstream sections of the Fangshan River, and the upstream section of the Luofu River had measured flows that did not meet the estimated ecological baseflow requirements. Considering the geographical distribution of the rivers, those failing to meet ecological baseflow thresholds or exhibiting interrupted flow were mostly concentrated in the eastern part of the region. Field investigation further indicated that many rivers in the east were intermittent, typically containing flow only after significant rainfall events.

4.3 Strategies for Maintaining Ecological Baseflow in the Northern Qinling Mountains

4.3.1 Analysis of Ecological Status of Rivers in the Northern Qinling Mountains

The northern piedmont of the Qinling Mountains serve as the primary water source region for the Guanzhong area, with numerous rivers originating there. The rivers flowing into the Wei River from the northern Qinling Mountains account for 61% of the total water resources in Guanzhong (Shaanxi Provincial Development and Reform Commission and Financial Department of Shaanxi Province 2018). In the upper and middle reaches of these rivers, several hydraulic engineering projects have been constructed. Notably, the Shitou River Basin's large-scale Shitou Reservoir is a key water supply project for major cities in Guanzhong, including Baoji, Xi'an, Xianyang, and Yangling, as well as for agricultural irrigation in the Qishan and Meixian areas; the reservoir is currently operating well. In the Fayu River Basin, the Diaoyutai Reservoir primarily serves tourism and sightseeing purposes. The periodic storage and release of water from these reservoirs significantly impact the living environments of aquatic organisms. Changes in hydrological conditions and interruptions in longitudinal river connectivity have adversely affected local biodiversity to some extent. In the middle and lower reaches of regional tributaries, rapid economic and population growth, accelerated urbanization, and concentrated human activities such as animal husbandry, industrial production, farmland fertilization, and domestic sewage discharge over recent decades have led to significant pollution of river water bodies, causing some degree of ecological health degradation (Guo 2019). In the downstream reaches, due to slowed water flow, widened river channels, and an increased proportion of fine sand and silt substrate, the biodiversity index in this area shows a relative upward trend.

Based on long-term data of annual rainfall and runoff measured over multiple years in the basin, analysis shows that both the annual rainfall and runoff in the Ba River basin are decreasing. The reduction in rainfall is the primary factor driving the decline in annual runoff. However, since the decreasing trend in runoff is more pronounced than that of rainfall, other factors, such as human activities, also contribute to the reduction in annual runoff alongside the decrease in precipitation.

Field surveys of ten rivers in the Baoji area of the northern Qinling Mountains found no sections experiencing flow interruption. Most river waters were clear and transparent, except for somewhat turbid conditions observed in the mid to lower reaches of Dongcha River, lower reaches of Yinxian River, the lower reaches of Mawei River and Panxi River near their confluences with the Wei River. Among the control cross sections established in the region, river sections meeting ecological baseflow targets were mainly concentrated in Taiyinyu River, Yinxian River, Mawei River, and Dongcha River. Eight control sections did not meet target requirements, including two upstream tributaries of Qingjiang River, two upstream tributaries of Dongcha River and Xicha River in the upper reaches of Qingshui River, the upper reaches of Panxi River, the upper reaches of Fayu River, and its tributaries Yama River and Shierpangou. Field surveys in the Xi'an section of the northern Qinling Mountains revealed that only one small river downstream of the Shibianyu Reservoir experienced flow interruption, which local villagers reported began after the reservoir was built. Comparing the calculated ecological baseflow with measured discharges at control sections in the Xi'an section showed that river reaches failing to meet ecological baseflow requirements were mainly concentrated in upstream segments of certain rivers. In the Weinan section of the northern Qinling Mountains, the greatest number of flow interruptions was observed. During the investigation, seven rivers (Liuye River, Changjian River, Baimajian River, Shuiyu River, Yuxian River, Shidi River, and Gouyu River) had some sections experiencing interrupted flow.

Overall, the ecological environment of rivers in the northern Qinling Mountains is generally good; however, certain areas exhibit activities that harm river ecological functions. For example, valley mouths are heavily impacted by human activities, with local residents engaging in barbecue and picnic gatherings along riverbanks and improperly disposing of waste, which easily affects water quality in the upper reaches. Additionally, sand excavation was observed during the survey, which severely undermines river stability, threatens the safety of hydraulic infrastructure, and can lead to water shortages in the river channel.

4.3.1.1 Causes of Flow Interruption in Representative Rivers

Human civilization has historically settled near water, but under the current backdrop of climate change and water scarcity, many rivers can no longer be classified as perennial. They experience varying degrees of flow interruption throughout the year, and some rivers remain dry for extended periods. According to the 2021 scientific expedition on the northern Qinling Mountains, seven rivers were identified with interrupted flow. Among them, no interrupted rivers were found in the Baoji section; only one river, Shibianyu (a tributary), exhibited interrupted flow in the Xi'an section; and six rivers in the Weinan section, including Shidi River, Changjian River, Liuye River, Luowen River, Shuiyu River, and Baimajian River, showed flow interruptions. The northern Qinling area lies within a warm temperate semi-arid to semi-humid continental monsoon climate zone, characterized by cold and dry winters and hot, rainy summers. Precipitation is highly variable, with frequent droughts. Rainfall is unevenly distributed seasonally, mainly concentrated in summer and autumn, accounting for 70–80% of annual precipitation. The river flow measurements in this study were conducted in August, the rainy season; nevertheless, many rivers still experienced interrupted flow. The following describes the status of these interrupted rivers and analyzes the underlying causes of flow cessation.

1. During the survey, no dry river sections were found in the Baoji segment of the northern Qinling Mountains. In the Xi'an segment, a tributary downstream of the Shibianyu Reservoir was observed to be dry. In the Weinan segment, flow interruptions occurred in the Shidi River, Changjian River, Liuye River, Luowen River, Shuiyu River, and Baimajian River. This regional flow interruption is preliminarily attributed to the low precipitation in the Weinan area. According to data, precipitation at the front of the Qinling Mountains shows significant spatiotemporal variability, generally higher in the south and lower in the north. The 2018 Shaanxi Provincial Water Resources Bulletin precipitation isohyet map clearly shows that Weinan receives less rainfall than Baoji and Xi'an. Table 4.7 compares precipitation in Xi'an, Baoji, and Weinan over the recent 3 years (2017–2019) with the long-term average. In 2019, Weinan's annual average precipitation was 535.6 mm, 7.9% less than the long-term average of 581.6 mm (1956–2000). Table 4.7 shows the minimum annual rainfall in the last 3 years was 482.9 mm, occurring in Weinan. Moreover, Weinan's precipitation in both

Table 4.7 Annual average precipitation in Xi'an, Baoji, and Weinan (mm)

Year	Xi'an	Baoji	Weinan
2017	822.9	753.2	582.4
2018	728.1	726.1	482.9
2019	909.7	780.5	535.6
Long-term average precipitation (1956–2000)	740.8	692.3	581.6

Table 4.8 Total water resources of Xi'an, Baoji, and Weinan in 2019 (10^8 m^3)

Administrative region	Precipitation	Surface water resources	Groundwater resources	Total water resources
Xi'an	88.70	23.57	11.72	27.46
Baoji	141.52	31.39	16.10	35.46
Weinan	61.67	5.53	8.57	9.88

recent years and long-term averages is below 600 mm, while Xi'an and Baoji have had annual rainfall over 700 mm in the same period. Therefore, it can be concluded that the main reason for multiple river flow interruptions in the Weinan segment is the sparse precipitation and arid climate in that region. Rivers in the northern Qinling Mountains are primarily replenished by atmospheric precipitation, so when precipitation decreases, most rivers cannot maintain their flow, resulting in dry riverbeds.

2. Several of the intermittent rivers in Weinan, including Changjian River, Luowen River, Liuye River, Baimajian River, and Shidi River, are all located in Huazhou District. It is therefore hypothesized that the geological environment in this area may also contribute to the river discontinuity. According to hydrogeological data, the aquifer formations in Huazhou District consist of unconsolidated porous aquifers and fractured karst aquifers in magmatic rocks, both of which are characterized as having weak groundwater abundance. In contrast, the Baoji and Xi'an sections are dominated by unconsolidated porous aquifers with moderate to high groundwater abundance. This comparison indicates that an important reason for the higher frequency of river flow interruption in Huazhou District of Weinan is the local geological condition, which is less favorable for water storage and retention.
3. According to the 2019 Shaanxi Water Resources Bulletin, as shown in Table 4.8, the total water resources in Weinan, including precipitation, surface water resources, and groundwater resources, are significantly lower than those in Xi'an and Baoji. Specifically, the total water resources in Weinan are only about one-third of those in Xi'an and two-sevenths of those in Baoji. However, data on water usage in 2019 (Table 4.9) show that the volume of water used in Weinan for agricultural irrigation, forestry, animal husbandry, and fishery was much higher than that of the other two cities. In fact, the total annual water consumption in Weinan was comparable to that of Xi'an and more than twice that of Baoji.

Table 4.9 Water consumption in Xi'an, Baoji, and Weinan in 2019 (10^8 m^3)

Administrative region	Agricultural irrigation	Forestry and agriculture	Industry	Residential use	Urban public	Ecological environment	Water consumption	
							Total water consumption	Surface water consumption
Xi'an	4.57	0.94	4.45	4.35	1.85	2.64	18.80	8.75
Baoji	3.94	1.09	0.99	1.39	0.23	0.21	7.85	3.94
Weinan	8.99	2.02	1.10	1.39	0.28	0.36	14.14	8.69

The combination of low water availability and high water demand has resulted in an overall scarcity of water resources in Weinan, which is closely linked to the frequent river flow interruptions observed in the region.

4. Besides the reduction in basin precipitation, the interception by water conservancy projects and increased socio-economic water use are also important causes of streamflow cessation. Water conservancy projects, especially dams built in arid areas for water storage, often directly cause downstream sections to run dry. For example, in the Xi'an segment, the Shibianyu River has experienced many years of no water in the middle and lower reaches due to an upstream reservoir. In the Baoji segment, although the Mawei River has water in its upper and middle reaches, the riverbed dries up downstream, where artificial embankments have been constructed on both sides. Human modifications of natural rivers have affected the local climate, which in turn impacts river flow.

River flow cessation seriously affects the production and daily life of local people. Besides domestic water use, industrial and agricultural water supplies mainly rely on surface water or shallow groundwater, so river drying has a significant impact on local communities. To address the problem of river drying caused by low rainfall, water conservancy projects, and human activities in the study area, agricultural structure adjustments can be made to reduce water consumption and promote water-saving agriculture.

4.3.1.2 Measures for Sustaining River Ecological Flow

Ensuring ecological flow is a systematic engineering endeavor that must adhere to dual objectives: addressing aquatic ecosystem issues while meeting water resource management requirements. Overall, the strategies for safeguarding ecological flow in rivers and lakes must maintain the integrity of river systems, balance water resources with ecological and socio-economic considerations, and holistically account for factors such as natural water endowment conditions, utilization patterns and demands, as well as ecological conservation requirements. The approach should combine necessity with feasibility to establish practical and effective ecological flow management solutions. The overarching framework for river ecological flow protection is as follows:

1. Balancing the relationship between humans and nature. Ecological flow protection measures should follow the principle of harmonious coexistence between humans and nature, ensuring both reasonable water demands for human socioeconomic development and the health of river systems. For irrational human activities and water use demands in socioeconomic development, adjustments and controls should be implemented based on the carrying capacity of basin and regional water resources and aquatic environments as the upper limit, with full efforts to guarantee ecological flows that maintain basic connectivity, basic morphology, basic environmental capacity, and basic habitats in river systems.

2. Respecting the natural principles of rivers and the growth characteristics of aquatic organisms. Ecological flow protection measures should be based on the natural laws governing rivers, honoring their interannual and intra-annual flow variation patterns, while comprehensively considering the diverse requirements of aquatic organisms and other factors to formulate appropriate protection measures and implementation processes.
3. Exploring reliable approaches for ecological flow protection based on historical and current conditions. Ecological flow protection measures must fully account for historical factors such as river water resource development, human settlements, and socioeconomic patterns to rationally determine river ecological conservation objectives and standards. For rivers with intensive development utilization and significant human impacts, practical circumstances should be considered to establish phased ecological flow protection strategies, progressively achieving coordinated development among water resources, socioeconomic systems, and ecological environments.
4. Emphasizing the systematic and long-term effectiveness of ecological flow protection measures. Given the extensive scope of ecological flow protection, measures should be ecosystem-based and implemented systematically for upstream-downstream relationships, mainstem-tributary systems, and riverbank-aquatic zone interactions. As ecological flow protection constitutes a long-term endeavor, the sustained effectiveness of all countermeasures must be thoroughly considered.
5. Highlighting the adaptive nature of ecological flow protection measures. As human-defined ecological flows evolve with deepening understanding of aquatic ecosystems and organismal behaviors, all protective measures must undergo adaptive adjustments aligned with stage-specific ecological conservation objectives across different implementation periods.

The Qinling Mountains are the most important ecological security barrier in central China, with abundant water resources being a core element and a prominent feature of the Qinling ecosystem. Among these, the northern piedmont of the Qinling Mountains hold 3.5 billion m^3 of water resources. The Wei River, the largest tributary of the Yellow River, serves as the primary water source for various counties in the northern Qinling area as well as major urban clusters including Xi'an, Baoji, and Weinan, nurturing the famous "Eight Rivers Surround Chang'an" landscape. Ecological flow is a crucial indicator for maintaining river ecological functions and regulating water resource development; it is a fundamental element for sustaining the ecosystems of rivers, lakes, and reservoirs. Therefore, maintaining river ecological flow at a reasonable level is a good measure for implementing ecological civilization construction in the current period. Shaanxi Province has introduced a series of protective measures targeting the Qinling ecological environment, scientifically established ecological flow benchmarks, and formulated ecological flow scheduling plans, laying a solid foundation for maintaining and improving the water ecology of the watershed. The following measures and suggestions are provided.

1. Strictly enforce relevant laws and policies
   Shaanxi provincial departments have successively issued laws and regulations such as the "Overall Plan for Ecological Protection of Shaanxi Qinling", "Water Ecological Management Plan for the Northern Piedmont of Shaanxi Qinling", "Qinling Ecological Environment Protection Regulations of Shaanxi Province", "Comprehensive Plan for the Wei River Basin", "Eight Waters Nourish Xi'an Plan", "Qinling Ecological Environment Protection Regulations of Xi'an", and "Qinling Ecological Environment Protection Plan of Xi'an". Local authorities need to earnestly implement these policies while promoting public awareness to enhance ecological protection consciousness, thereby establishing a long-term mechanism for Qinling ecological environment protection led by the government and involving multiple stakeholders.
   Specifically targeting rivers such as the Shidi River in Weinan and the upper reaches of the Ba River in Xi'an, field researchers discovered that villagers were engaged in sand mining within the upstream river channels. This activity severely damages the natural river morphology and causes irreversible harm to the river's ecological environment. It is recommended that relevant authorities strengthen public awareness campaigns to improve community consciousness about ecological protection, and regularly organize patrols to firmly prevent behaviors that harm the Qinling ecological environment.
2. Reasonably plan watershed water resources and strictly enforce water extraction permit management
   Scientifically plan watershed water resources by following the total water use control schemes set by higher authorities, specifying ecological water requirements for each control section of the watershed at different time periods. During the development and utilization of watershed water sources, river ecological flow and ecological water levels must be taken into account. Strict management of water extraction permits is essential, especially for water use in construction projects. Continuous monitoring and inspection of water usage should be conducted to ensure that river ecological flow remains at a stable level.
   Especially for the Weinan section of the northern Qinling foothills, where production and domestic water demand is high but overall water resources are relatively limited, it is crucial to strengthen the rational use of surface water and groundwater to avoid excessive groundwater extraction. Over-extraction reduces groundwater recharge to rivers, causing a decrease in river flow that fails to meet ecological baseflow requirements. At the same time, efforts should be increased to promote water-saving agricultural irrigation practices to reduce water waste.
3. Strengthen flow monitoring and management
   Strengthening flow monitoring and management, and conducting thorough assessments of river ecological flow, are crucial measures for protecting and maintaining the ecological environment of the northern Qinling region. It is necessary to enhance the unified planning and management of monitoring stations, define clear monitoring indicators for each site, and establish a comprehensive watershed monitoring network to improve the timeliness and relevance of monitoring efforts within the basin. The use of advanced technologies should be pro-

moted to build a robust information-sharing platform, allowing for water quantity assessments based on unified and fair evaluation standards, thereby better achieving integrated water resource management across the watershed.

Currently, hydrological stations in the northern Qinling region are only established on some medium and large rivers, while numerous small rivers lack monitoring, making it difficult to adjust and implement measures based on actual river flow conditions to ensure the maintenance of ecological baseflow. It is recommended to increase flow monitoring, especially for rivers with upstream reservoirs, such as Shitou River and Shibianyu River. Flow monitoring should be strengthened downstream of reservoirs, and reservoir releases should be carried out when necessary to ensure river flows meet ecological baseflow requirements, thereby maintaining the basic ecological functions of the river channels.

4. River early warning and allocation management

   Based on the water resource conditions and socioeconomic development levels of different watersheds, dynamic ecological water use management targets should be established. For example, in dry years, ecological water use can be moderately reduced, while in normal years, the ecological water needs of rivers should be met as fully as possible. A river ecological water scheduling and early warning flow system should be set up. Different warning levels correspond to the degree to which minimum ecological flow and suitable ecological flow are met: the suitable ecological flow level is set as a yellow warning flow, and the minimum ecological flow level as a red warning flow. Under water scarcity conditions, the minimum ecological water requirement is established as the ecological crisis warning flow (water level) of the river, serving as the cutoff or drying alert line. When the river flow falls below this line, there is a high risk of flow cessation. This system is analogous to flood warning levels, where exceeding the flood warning line does not necessarily cause flooding but prompts unconventional management measures.

   Establish a river ecological water crisis management mechanism. The minimum ecological water requirement represents the bottom line for maintaining the health of the river ecosystem. Under any circumstances, efforts must be made to keep the river flow above this critical survival threshold. The minimum ecological water requirement is most vulnerable during dry seasons; if flow falls below this minimum ecological flow, the river enters a crisis state. When river flow during the dry season drops below the warning line (minimum ecological flow), water withdrawal management must enter an emergency mode. Below the warning line, a risk defense line for flow cessation (or drying) is set, and ecological water allocation indicators should be moderately reduced, maintaining pulsations between the defense line and the warning line to prevent complete flow stoppage. To resolve conflicts between economic water use (which typically peaks during dry seasons) and ecological water needs at this time, a crisis management system should be established to restrict water withdrawals. From a value perspective, river runoff below the red warning line is the "life water" of the river, whose value far exceeds that of normal water levels and should be subject to "high-cost" restrictions.

4.4 Summary

This study conducted a comprehensive investigation and analysis of the ecological status and ecological baseflow of medium- and small-sized rivers in the northern piedmont of the Qinling Mountains through data collection, field surveys, resident interviews, flow measurements, cross-sectional measurements, and laboratory analyses. The main conclusions are as follows:

1. The ecological status of 55 rivers in the northern piedmont of the Qinling Mountains was investigated and analyzed. Cross-sections were established to measure flow and identify the basic characteristics of river segments. The results show that the ecological environment of medium- and small-sized rivers in this area is generally good; however, activities harmful to the river ecology, such as sand excavation and dumping of construction waste, were observed. Intermittent flow or dry river conditions were found in rivers including the Liuyeye River, Changjian River, Baimajian River, Shuiyu River, Yuxian River, Shiti River, and Gouyu River. Regionally, river dry-up occurrences are mainly concentrated in the Weinan area. Surface water resources are unevenly distributed across the region, and flow rates vary significantly among river segments, ranging from 0.005 to 8.193 m^3/s. The river with the highest flow rate is the Taiyiyu River (measured after significant rainfall in the area), while the lowest flow was observed in the upper reaches of the Shuiyu River.
2. Based on a thorough understanding of the flow conditions of different segments of medium- and small-sized rivers in the study area, combined with the cross-sectional morphology of the rivers, the ecological baseflow for different river segments was calculated using the wet perimeter method, and compared with actual measured flows. The analysis shows that some cross-sections in the upper tributaries of the Qingjiang River, upper tributaries of the Qingshui River, upper tributaries of the Fayu River, and Tongyu River in the Baoji area; certain cross-sections of the Ziwuyu River, Baolongyu River, Tianziyu River, and Dayu River in the Xi'an area; as well as cross-sections in the upper reaches of the Yuxian River, upper and middle reaches of the Fangshan River, and upper reaches of the Luofu River in the Weinan area have measured flows that do not meet the estimated ecological baseflow. It was also found that cross-sections failing to meet the estimated ecological baseflow are mainly concentrated in the upper reaches of the rivers.
3. Through comparative analysis of the current flow conditions and ecological baseflows of rivers in the northern Qinling region, combined with regional hydrogeological and meteorological data as well as ecological environment surveys, the causes of river drying in the area were analyzed. River drying is closely related to the spatial unevenness of regional rainfall distribution and may also be influenced by local hydrogeological conditions. The Weinan area, near mountainous regions, mainly consists of metamorphic aquifers with weak water storage capacity, resulting in limited groundwater recharge to rivers. Rainfall mainly produces direct surface runoff, causing many rivers in the northern Qinling to exhibit seasonal flow characteristics, especially in the Weinan area. Additionally, human activities such as reservoirs and other water interception facilities also impact river flow.

References

Gippel CJ, Stewardson MJ (1998) Use of wetted perimeter in defining minimum environmental flows. Regul Rivers: Res Manage 14(1):53–67. https://doi.org/10.1002/(SICI)1099-1646(199801/02)14:13.0.CO;2-Z

Guo J (2019) Ecological restoration and numerical simulation of water environment in Bahe River Basin. Master thesis, Xi'an University of Technology (In Chinese with English Abstract)

Huang Q (2013) Hydraulics tutorial, 4th edn. Southwest Jiaotong University Press, Chengdu. (In Chinese)

Ji L (2006) Study on hydraulic methods calculating instream baseflow requirement. Master thesis, Northwest A&F University (In Chinese with English Abstract)

Ji L, Liu S, Lv H, Men B (2006) Theoretical study of minimum instream flow requirement using the wetted perimeter method. J Northwest A&F Univ (Nat Sci Ed) 34(2):124–130. (In Chinese with English Abstract)

Ji L, Liu S, Wang X (2010) Wetted perimeter approach to estimate instream flow requirements: a case study in Luanhe water system. Prog Geogr 29(3):287–291. (In Chinese with English Abstract)

Liu S, Mo X, Xia J et al (2006) Uncertainty analysis in estimating the minimum ecological instream flow requirements via wetted perimeter method: curvature technique or slope technique. Acta Geograph Sin 61(3):273–281. (In Chinese with English Abstract)

Shaanxi Provincial Development and Reform Commission, and Financial Department of Shaanxi Province (2018) Integrated management of the Qinling mountain ecosystem. China Development Press, Beijing. (In Chinese)

Xu Z, Wu W, Yu S (2016) Ecological baseflow: progress and challenge. J Hydroelectr Eng 35(4):1–11. (In Chinese with English Abstract)

Xu Z, Li P, Hou X (2019) Theoretical basis and estimation method for ecological base-flow. Yellow River 41(10):119–127. (In Chinese with English Abstract)

Zhang Y, Wei J, Huang S et al (2021) Research on ecological flow of small and medium-sized rivers in Jinna Mountain area based on wet cycle. Yellow River 43:1–5. (In Chinese with English Abstract)

Chapter 5
Surface Water Quality in the Northern Qinling Mountains

Abstract This chapter investigates the surface water quality in the northern Qinling Mountains. Through systematic field surveys, 156 surface water samples were collected from rivers and reservoirs across the region. The chapter employed multiple analytical methods, including hydrochemical characterization (Piper and Gibbs diagrams), correlation analysis, multivariate statistics, and stable isotope techniques, to elucidate the formation mechanisms and influencing factors of surface water quality. Assessments were conducted for drinking water, irrigation, and industrial uses. The chapter also identifies ecological challenges such as water scarcity, pollution, and habitat degradation, proposing measures for sustainable water resource management. This chapter provides a scientific foundation for ecological protection policies and supports the implementation of national strategies for environmental conservation in the Qinling region.

Keywords Water quality assessment · Hydrochemistry · Drinking water quality · Irrigation water quality · Industrial water quality · Water pollution

5.1 Introduction

Water resources are irreplaceable and essential, serving as the source of all life and closely tied to the environment on which humanity depends (Stringer et al. 2021). At present, environmental pollution has become increasingly severe worldwide, drawing growing public attention to environmental protection, with numerous public crises triggered by water pollution. Currently, approximately 40% of the world's rivers suffer from varying degrees of pollution, with a continuing upward trend. Due to unsustainable environmental practices, the deterioration of river water quality has emerged as a major environmental issue. River water quality is influenced by a complex interplay of human activities and natural factors at both river and watershed scales (Uddin et al. 2021). Natural factors primarily include topography and

P. Li et al., *Water Resources Research in Northern Qinling Mountains, China*, Springer Hydrogeology, https://doi.org/10.1007/978-3-032-09395-0_5

climate change, while anthropogenic factors involve socioeconomic activities, livestock and poultry production, and land use, among which human influences tend to be more significant (Chen 2014). For the prevention and control of river water pollution, water quality assessment serves as a fundamental basis. It involves quantitatively evaluating the quality of water bodies using extensive monitoring data (Salerno et al. 2018; Yan et al. 2022). Such evaluations help determine the degree of water pollution, provide early warnings for drinking water safety, and offer a scientific basis for formulating pollution control strategies. A healthy watershed aquatic ecosystem is essential for maintaining riverine ecological security, preserving biodiversity, and ensuring the sustainable development of watershed socioeconomic systems (Gao 2022).

As the geographical boundary between northern and southern China, the Qinling Mountains serve not only as an important ecological barrier but also as a key national strategic water conservation area and a biodiversity hotspot. The vast river network on the northern slope of the Qinling nourishes the Guanzhong Basin, directly influencing regional drinking water safety and ecological health (Ma et al. 2011; Wang et al. 2023). In recent years, with the advancement of major national strategies such as the "Ecological Conservation and High-Quality Development of the Yellow River Basin" and the "Middle Route of the South-to-North Water Diversion Project," the ecological value and strategic significance of the Qinling region have become increasingly prominent. Under mounting pressures from resource development, urban–rural construction, and climate change, the water quality of rivers in the Qinling region faces the risks of pollution and degradation, potentially threatening regional water security and ecological balance. Therefore, a systematic study of water quality changes in rivers across the Qinling region is not only crucial for maintaining the stability of watershed ecosystems but also provides a scientific foundation for the implementation of national ecological civilization and green development policies. By scientifically monitoring and evaluating the water quality of major rivers in Shaanxi Province, analyzing the concentrations of key pollutants and their deviations from national water quality standards, and accurately determining water quality classifications, this research offers essential support for ecological protection of the Qinling Mountains, sustainable use of water resources, and the implementation of national strategies. Furthermore, it provides a theoretical basis for water pollution control, contributing significantly to the coordinated advancement of economic development and environmental protection.

5.2 Sample Collection and Analysis

This study collected a total of 156 surface water samples from the northern Qinling Mountains, including 147 river water samples and nine reservoir water samples, covering approximately 50 rivers and nine reservoirs in the region. Water samples were collected at upstream, midstream, and downstream sections of each river. All sampling bottles were rinsed four to five times with the sample water prior to collection. The samples were used to analyze major cations and anions, heavy metals, and isotopes. For heavy metal analysis, two drops of nitric acid were added to the samples to ensure accuracy, and isotope samples were collected to prevent air contamination. All samples were transported back to the laboratory for analysis as quickly as possible. Water quality samples were collected following the "*Technical Specifications for Monitoring Surface Water and Wastewater*" (HJ/T 91-2002). Sampling and preservation methods for various analyses referenced the "*Technical Regulations on Preservation and Management of Water Quality Sampling Samples*" (HJ 493-2009). Proper preservation and transportation measures were maintained throughout the sampling process. Seven samples were additionally tested for microbiological parameters, including total bacterial counts and *Escherichia coli* numbers. Microbial samples were collected using sterile bottles, which were rinsed with sample water before collection, taken from the laboratory within 48 h prior to sampling, and returned to the laboratory for analysis within 48 h after sampling.

5.3 Surface Water Quality

5.3.1 Basic Characteristics of Surface Water Hydrochemistry

The statistical results of the main chemical indicators of surface water in the study area are shown in Table 5.1. Overall, the pH of surface water ranges from 6.98 to 9.46, and the concentrations of various ions exhibit relatively small variations. The concentration of Na^+ ranges from 1.46 to 134.00 mg/L. Among the cations, K^+ has the lowest concentration, ranging from 1.83 to 32.30 mg/L, with an average concentration of 4.81 mg/L, which is about 24% of the Na^+ content. This is mainly due to the weak mobility of K^+ determined by its biological activity, as organic matter from plants and animals can absorb K^+ from the water. The concentrations of Ca^{2+} and Mg^{2+} range from 16.00 to 130.00 mg/L and 0.49 to 37.70 mg/L, respectively, with averages of 38.94 mg/L and 9.51 mg/L. Among the major anions, Cl^- shows the smallest concentration range (0.40–185.00 mg/L), while HCO_3^- has the largest range (36.60–305.00 mg/L). SO_4^{2-} concentrations vary from a minimum of 4.80 mg/L to a maximum of 226.00 mg/L, with an average of 46.60 mg/L. The concentration order of major anions in the surface water is $HCO_3^- > SO_4^{2-} > Cl^-$, and for major cations, it is $Ca^{2+} > Na^+ > Mg^{2+} > K^+$. Total dissolved solids (TDS) range from 68.00 to 808.00 mg/L, meeting the quality standards for drinking water. Analysis of the main cations (Ca^{2+}, Na^+, Mg^{2+}, K^+) shows that Na^+ and K^+ concentrations generally increase from upstream to midstream to downstream, while Ca^{2+} and Mg^{2+} exhibit a pattern where concentrations are lowest upstream and highest midstream. For the main anions (HCO_3^-, SO_4^{2-}, Cl^-), SO_4^{2-} and Cl^- concentrations also increase from upstream to downstream, whereas HCO_3^- follows a similar pattern to Ca^{2+} and Mg^{2+}, with the lowest concentrations upstream and highest midstream. The concentrations of both major cations (Ca^{2+}, Na^+, Mg^{2+}, K^+) and major anions

Table 5.1 Statistical summary of major ions and chemical indicators (mg/L)

Locations	Statistics	pH	K^+	Na^+	Ca^{2+}	Mg^{2+}	Cl^-	SO_4^{2-}	HCO_3^-	CO_3^{2-}	TDS
Upstream	Mean	8.07	2.97	4.80	34.11	5.07	4.00	29.66	92.85	0.69	150.37
	Min	7.09	1.88	1.46	16.00	1.22	0.40	9.61	36.60	0.00	68.00
	Max	8.85	6.69	24.70	76.20	19.40	17.00	110.00	250.00	18.00	356.00
Midstream	Mean	8.05	5.21	18.78	47.27	7.83	24.29	44.25	133.81	1.29	241.14
	Min	7.45	2.20	1.77	18.00	0.49	0.70	9.61	54.90	0.00	80.00
	Max	8.82	26.50	119.00	130.00	27.90	179.00	226.00	299.00	12.00	808.00
Downstream	Mean	7.92	7.21	24.92	41.80	7.59	27.79	46.03	128.97	0.82	243.64
	Min	6.98	2.31	1.69	18.00	1.22	1.00	9.61	36.60	0.00	92.00
	Max	9.46	32.30	134.00	120.00	21.90	185.00	183.00	275.00	24.00	724.00
Wei River	Mean	7.94	5.38	49.76	45.05	15.13	49.69	79.22	159.94	0.00	351.14
	Min	7.04	2.47	2.05	24.00	1.22	1.00	4.80	73.20	0.00	80.00
	Max	8.29	10.80	115.00	110.00	34.00	121.00	192.00	244.00	0.00	716.00
Reservoirs	Mean	8.30	3.29	28.29	26.49	11.94	13.77	33.82	136.29	9.33	212.00
	Min	7.42	1.83	1.77	18.00	1.22	0.60	9.61	48.80	0.00	84.00
	Max	9.06	4.99	82.10	38.10	37.70	44.00	81.70	305.00	48.00	420.00
Overall	Mean	8.06	4.81	25.31	38.94	9.51	23.91	46.60	130.37	2.43	239.66
	Min	6.98	1.83	1.46	16.00	0.49	0.40	4.80	36.60	0.00	68.00
	Max	9.46	32.30	134.00	130.00	37.70	185.00	226.00	305.00	48.00	808.00

(HCO_3^-, SO_4^{2-}, Cl^-) in the Wei River are higher than those in reservoir waters within the study area.

The pH of surface water in the upstream area ranges from 7.09 to 8.85, showing slight alkalinity, with relatively small variations in ion concentrations. The concentration of Na^+ ranges from 1.46 to 24.70 mg/L. Among cations, K^+ has the lowest concentration, with a maximum of 6.69 mg/L, about 30% of the Na^+ content. This is mainly due to the weak mobility of K^+ determined by its biological activity, as organic matter from plants and animals can absorb potassium ions from the water. The concentrations of Ca^{2+} and Mg^{2+} range from 16.00 to 76.20 mg/L and 1.22 to 19.40 mg/L, respectively, with averages of 34.11 mg/L and 5.07 mg/L. Among the major anions, Cl^- has the smallest concentration range (0.40–17.00 mg/L), while HCO_3^- has the largest range (36.60–250.00 mg/L). SO_4^{2-} concentrations vary from a minimum of 9.61 mg/L to a maximum of 110.00 mg/L, with an average of 29.66 mg/L. In the upstream area, the concentration order of major anions is $HCO_3^- > SO_4^{2-} > Cl^-$, and for major cations, it is $Ca^{2+} > Mg^{2+} > Na^+ > K^+$. TDS values range from 68.00 to 356.00 mg/L, meeting the quality standards for drinking water.

The pH of surface water in the midstream area ranges from 7.45 to 8.82, exhibiting slight alkalinity. Among cations, K^+ has the lowest concentration, with an average of 5.21 mg/L. Na^+ concentration varies between 1.77 and 119.00 mg/L. The concentrations of Ca^{2+} and Mg^{2+} range from 18.00 to 130.00 mg/L and 0.49 to 27.90 mg/L, respectively, with averages of 47.27 mg/L and 7.83 mg/L. Among the major anions, Cl^- and HCO_3^- concentrations range from 0.40 to 17.00 mg/L and 36.60 to 250.00 mg/L, respectively. SO_4^{2-} concentrations range from a minimum of 9.61 mg/L to a maximum of 110.00 mg/L, with an average of 29.66 mg/L. The concentration order of major anions in the area is $HCO_3^- > SO_4^{2-} > Cl^-$, and for major cations, it is $Ca^{2+} > Na^+ > Mg^{2+} > K^+$. TDS values range from 80.00 to 808.00 mg/L, meeting the quality standards for drinking water.

The pH of surface water in the downstream area ranges from 6.98 to 9.46. The concentration of Na^+ varies from 1.69 to 134.00 mg/L, and K^+ ranges from 2.31 to 32.30 mg/L. Ca^{2+} and Mg^{2+} concentrations range from 18.00 to 120.00 mg/L and 1.22 to 21.90 mg/L, with average values of 41.80 mg/L and 7.59 mg/L, respectively. Among the major anions, Cl^- and HCO_3^- concentrations vary from 1.00 to 185.00 mg/L and 36.60 to 275.00 mg/L, respectively. SO_4^{2-} concentration ranges from a minimum of 9.61 mg/L to a maximum of 183.00 mg/L, with an average of 46.03 mg/L. The concentration order of major anions in the downstream area is $HCO_3^- > SO_4^{2-} > Cl^-$, and for major cations, it is $Ca^{2+} > Na^+ > Mg^{2+} > K^+$. TDS values range from 92.00 to 724.00 mg/L, meeting the quality standards for drinking water.

The pH of the Wei River ranges from 7.04 to 8.29. The concentration of Na^+ varies between 2.05 and 115.00 mg/L, while K^+ ranges from 2.31 to 32.30 mg/L. The concentrations of Ca^{2+} and Mg^{2+} range from 24.00 to 110.00 mg/L and 1.22 to 34.00 mg/L, with averages of 45.05 mg/L and 15.13 mg/L, respectively. Among the major anions, HCO_3^- and SO_4^{2-} concentrations vary from 73.20 to 244.00 mg/L and 4.80 to 192.00 mg/L, respectively. Cl^- concentration ranges from a minimum of

1.00 mg/L to a maximum of 121.00 mg/L, with an average of 49.69 mg/L. In the downstream surface water of the study area, the major anion concentration order is HCO_3^- > SO_4^{2-} > Cl^-, and the major cation concentration order is Ca^{2+} > Na^+ > Mg^{2+} > K^+. TDS values range from 92.00 to 724.00 mg/L, meeting drinking water quality standards.

A total of 9 reservoir water samples were collected, including the You River Reservoir, Ling River Reservoir, Sanzhang Reservoir, Jinshan Reservoir, Diaoyutai Reservoir, Panxi River Reservoir, Tongyu River Reservoir, Shitou River Reservoir, and Xiluoyu Reservoir. The pH ranged from 7.42 to 9.06, with the lowest pH at Diaoyutai Reservoir and the highest at Ling River Reservoir. Among the major cations, Na^+, K^+, and Mg^{2+} concentrations had their minimum and maximum values in Shitou River Reservoir and Sanzhang Reservoir, respectively. The concentration ranges for Na^+, K^+, and Mg^{2+} were 1.77–82.10 mg/L, 1.83–4.99 mg/L, and 1.22–37.70 mg/L. The lowest Ca^{2+} concentration was 18.00 mg/L in Sanzhang Reservoir, and the highest was 38.10 mg/L in Shitou River Reservoir. Among the major anions, HCO_3^- had the widest concentration range (48.80–305.00 mg/L), with Panxi River Reservoir having 48.80 mg/L and Sanzhang Reservoir 305.00 mg/L. SO_4^{2-} concentrations ranged from 9.61 to 81.70 mg/L, with Shitou River Reservoir at the low end and You River Reservoir at the high end. Cl^- concentrations ranged from 0.6 mg/L at Xiluoyu Reservoir to 44 mg/L at Sanzhang Reservoir. TDS ranged from 84.00 to 420.00 mg/L, corresponding to Shitou River Reservoir and Sanzhang Reservoir, respectively.

Based on the water quality test results of surface water samples, the major hydrochemical types were analyzed. The overall chemical types for water samples collected are primarily HCO_3–Ca·Mg, SO_4·Cl–Na, and SO_4·Cl–Ca·Mg types. The water chemistry types are relatively simple. However, in different reaches of the rivers, the hydrochemical types diverse. The distribution of surface water chemical types in the study area is shown in Fig. 5.1.

The surface water chemical types in the upstream area of the study region are HCO_3–Ca·Mg type and SO_4·Cl–Ca·Mg type. The midstream area surface water chemical type is mainly HCO_3–Ca·Mg type. The downstream area surface water chemical types include HCO_3–Ca·Mg type and SO_4·Cl–Na type. The water chemistry types in the Wei River basin are HCO_3–Ca·Mg type and SO_4·Cl–Ca·Mg type. The water chemical type of the You River Reservoir is SO_4·Cl–Na type, while the Ling River Reservoir, Sanzhang Reservoir, Jinshan Reservoir, Diaoyutai Reservoir, Panxi River Reservoir, Tongyu River Reservoir, Shitou River Reservoir, and Xiluoyu Reservoir all have water of the HCO_3–Ca·Mg type.

5.3.2 *Evaluation of Surface Water Environmental Quality*

5.3.2.1 Drinking Water Quality Assessment

As drinking water, it must meet the requirements of being physiologically acceptable and harmless to human health. Therefore, the evaluation should consider the physical properties of the water, dissolved common salts, and also pay attention to

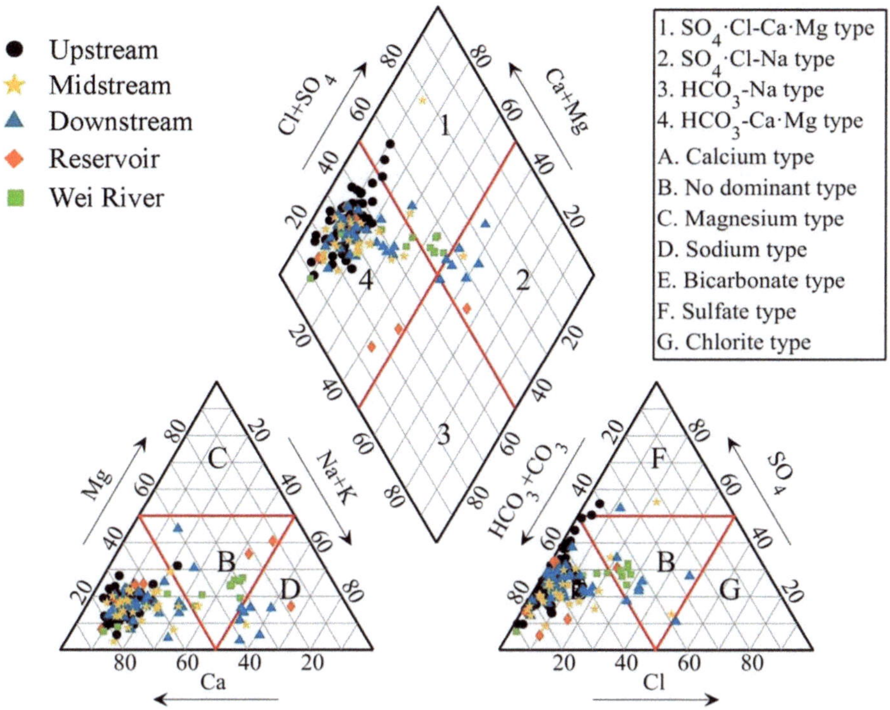

Fig. 5.1 Piper diagram of surface water hydrochemical types

whether the surface water is contaminated by bacteria and toxins. In this study, a total of 156 water samples (including 1 duplicate sample) were collected from the northern piedmont of the Qinling Mountains, including 9 reservoir samples, 15 samples from the Wei River, and 132 samples from other rivers. The testing items included 28 parameters such as color, turbidity, pH, various anions and cations, mineralization, total hardness, and other routine ions, as well as toxicological indicators such as fluoride, hexavalent chromium, lead, and microbiological indicators. For drinking water quality assessment, 19 parameters were selected, including sensory and general chemical indicators such as color, turbidity, TDS, total hardness, pH, chloride, sulfate, iron, manganese, chemical oxygen demand (COD), ammonia nitrogen, sodium; toxicological indicators including nitrate, nitrite, fluoride, hexavalent chromium, lead; and microbiological indicators such as total colony count and total coliform. The current status of drinking water quality for major river mouths in the northern piedmont of the Qinling was evaluated. The surface water quality evaluation standards are shown in Table 5.2, and the evaluation results are presented in Table 5.3. Grade I indicates excellent water quality, grade II suggests good quality, while grade III denotes medium water quality. Water falls within grades I to III are considered suitable for drinking. Grade IV indicates poor quality and grade V suggests very poor quality, both of which are not suitable for drinking.

According to the domestic drinking water quality evaluation standards, 155 surface water samples collected from the northern piedmont of the Qinling Mountains

Table 5.2 Surface water quality evaluation standards

Parameter	Unit	I	II	III	IV	V
Color	/	≤5	≤5	≤15	≤25	>25
Turbidity	NTU	≤3	≤3	≤3	≤10	>10
Sodium	mg/L	≤100	≤150	≤200	≤400	>400
Ammonia nitrogen (as N)	mg/L	≤0.02	≤0.1	≤0.5	≤1.5	>1.5
Iron	mg/L	≤0.1	≤0.2	≤0.3	≤2	>2
Chloride	mg/L	≤50	≤150	≤250	≤350	>350
Sulfate	mg/L	≤50	≤150	≤250	≤350	>350
Nitrate (as N)	mg/L	≤2	≤5	≤20	≤30	>30
Nitrite (as N)	mg/L	≤0.01	≤0.1	≤1	≤4.8	>4.8
TDS	mg/L	≤300	≤500	≤1000	≤2000	>2000
TH	mg/L	≤150	≤300	≤450	≤650	>650
COD	mg/L	≤15	≤15	≤20	≤30	>40
Fluoride	mg/L	≤1	≤1	≤1	≤2	>2
Hexavalent chromium (Cr^{6+})	mg/L	≤0.005	≤0.01	≤0.05	≤0.1	>0.1
Lead	mg/L	≤0.0025	≤0.005	≤0.01	≤0.1	>0.1
Manganese	mg/L	≤0.025	≤0.05	≤0.1	≤1.5	>1.5
Total bacterial count	CFU/mL	≤100	≤100	≤100	≤1000	>1000
Total coliform bacteria	MPN/100 mL	Not detectable	Not detectable	Not detectable	≤100	>100
pH	6.5 ≤ pH ≤ 8.5				5.5 ≤ pH < 6.5 8.5 < pH ≤ 9.0	pH <5.5 or pH >9

Table 5.3 Surface water quality evaluation results

Sample ID	Water quality grade	Sample ID	Water quality grade	Sample ID	Water quality grade
QL1-01	II	QL1-53	III	QL2-32	II
QL1-02	II	QL1-54	II	QL2-33	II
QL1-03	II	QL1-55	V	QL2-34	II
QL1-04	II	QL1-56	II	QL2-36	III
QL1-05	II	QL1-57	II	QL2-37	II
QL1-06	IV	QL1-58	III	QL2-38	II
QL1-07	V	QL1-59	III	QL2-39	IV
QL1-08	III	QL1-60	IV	QL2-41	III
QL1-09	V	QL1-61	II	QL2-42	IV
QL1-10	V	QL1-62	II	QL2-43	III
QL1-11	IV	QL1-63	IV	QL2-44	IV
QL1-12	IV	QL1-64	IV	QL2-45	II
QL1-13	IV	QL1-65	III	QL2-46	III
QL1-14	III	QL1-66	V	QL2-47	V
QL1-15	V	QL1-67	IV	QL2-48	V
QL1-16	IV	QL1-68	IV	QL2-49	II
QL1-17	II	QL1-69	II	QL2-50	II
QL1-18	III	QL1-70	V	QL2-51	II
QL1-19	II	QL1-71	IV	QL2-52	III
QL1-20	II	QL1-72	V	QL2-53	IV
QL1-21	II	QL1-73	V	QL2-54	III
QL1-22	IV	QL1-74	V	QL2-55	V
QL1-23	II	QL2-01	II	QL2-56	III
QL1-24	II	QL2-02	III	QL2-57	III
QL1-25	II	QL2-03	III	QL2-58	V
QL1-26	III	QL2-04	V	QL2-59	II
QL1-27	IV	QL2-05	V	QL2-60	II
QL1-28	III	QL2-06	V	QL2-61	IV
QL1-29	III	QL2-08	II	QL2-62	III
QL1-30	III	QL2-09	V	QL2-63	III
QL1-31	II	QL2-10	V	QL2-64	II
QL1-32	II	QL2-11	IV	QL2-65	II
QL1-33	II	QL2-12	II	QL2-66	III
QL1-34	V	QL2-13	IV	QL2-67	IV
QL1-35	V	QL2-14	V	QL2-68	IV
QL1-36	II	QL2-15	II	QL2-69	IV
QL1-37	II	QL2-16	III	QL2-70	III
QL1-38	II	QL2-17	IV	QL2-71	IV
QL1-39	II	QL2-18	IV	QL2-72	IV
QL1-40	II	QL2-19	IV	QL2-73	III
QL1-41	II	QL2-20	IV	QL2-74	III

(continued)

Table 5.3 (continued)

Sample ID	Water quality grade	Sample ID	Water quality grade	Sample ID	Water quality grade
QL1-42	IV	QL2-21	III	QL2-75	V
QL1-43	III	QL2-22	IV	QL2-76	V
QL1-44	III	QL2-23	IV	QL2-77	V
QL1-45	II	QL2-24	IV	QL2-78	V
QL1-46	II	QL2-25	II	QL2-79	V
QL1-47	V	QL2-26	V	QL2-80	IV
QL1-48	V	QL2-27	V	QL2-81	V
QL1-49	II	QL2-28	V	QL2-82	V
QL1-50	II	QL2-29	II	QL2-83	III
QL1-51	IV	QL2-30	V	QL2-84	V
QL1-52	II	QL2-31	II		

were classified by water quality grades. Among them, there were no samples meeting Class I standards. Class II water quality accounted for 52 samples (33.55% of the total), Class III for 32 samples (20.64%), Class IV for 35 samples (22.58%), and Class V for 36 samples (23.23%). Water of Class I, II, and III which are considered suitable for human consumption made up 54.19 of the total samples. The primary indicators affecting water quality evaluation are turbidity, NH_4^+, COD, Fe, Mn, and F^-.

5.3.2.2 Irrigation Water Quality Assessment

The quality of irrigation water significantly affects soil structure and plant growth, making it essential to evaluate irrigation water quality. Currently, the main methods or indicators used for assessing agricultural irrigation water quality include Sodium Adsorption Ratio (SAR), Residual Sodium Carbonate (RSC), Sodium Percentage (%Na), and Permeability Index (PI). The specific equations are shown as Eqs. (5.1), (5.2), (5.3) and (5.4). SAR represents the soil's ability to adsorb sodium ions from irrigation water; a higher SAR indicates a stronger capacity for sodium adsorption, which in turn affects the plant roots' ability to absorb water. RSC indicates the potential for alkalinity hazards caused by irrigation water, with larger RSC values suggesting a higher likelihood of such hazards. %Na also reflects the degree of alkalinity hazard posed by irrigation water. The PI is influenced by long-term irrigation as well as the content of Na, Ca, Mg, and bicarbonate in the soil. All of these indicators reflect the potential for alkalinity hazards caused by irrigation water; however, salinity hazards should also be taken into account when evaluating irrigation water quality. Salinity is commonly represented by electrical conductivity (EC). The calculation formulas and evaluation standards for these indicators are presented in the following table (Table 5.4).

Table 5.4 Irrigation water quality evaluation standards

SAR	EC	Irrigation water quality	RSC	%Na	PI	Irrigation water quality
<10	<250	Excellent	<1.25	<30	>75	Suitable
10–18	250–750	Good	1.25–2.5	30–60	25–75	Basically suitable
18–26	750–2250	Moderate	>2.5	>60	<25	Not suitable
>26	>2250	Poor				

$$SAR = \frac{Na^{+}}{\sqrt{\frac{Ca^{2+} + Mg^{2+}}{2}}} \tag{5.1}$$

$$RSC = \left(CO_3^{2-} + HCO_3^{-}\right) - \left(Ca^{2+} + Mg^{2+}\right) \tag{5.2}$$

$$\%Na = \frac{Na^{+} + K^{+}}{Ca^{2+} + Mg^{2+} + Na^{+} + K^{+}} \times 100\% \tag{5.3}$$

$$PI = \frac{Na^{+} + \sqrt{HCO_3^{-}}}{Ca^{2+} + Mg^{2+} + Na^{+}} \times 100\% \tag{5.4}$$

The SAR, RSC, %Na, and PI indicators all reflect the degree of alkali hazard caused by irrigation water (Table 5.5). According to the SAR results, all water samples were rated as excellent. Based on the RSC evaluation, only two samples were rated as basically suitable for irrigation, while the rest were suitable. According to the %Na evaluation, 19 samples were rated as basically suitable for irrigation, and the remainder were suitable. The PI evaluation showed eight samples as suitable for irrigation, with the others rated as basically suitable. From the above alkali hazard assessments, although there are slight differences among the evaluation indicators, overall the surface water in the study area is suitable for irrigation. According to the EC evaluation, out of 148 surface water samples, 76 were rated as excellent, accounting for 51.35% of the total samples; 63 were rated as good, accounting for 42.57%; and nine were rated as moderate, accounting for 6.08%. Surface water rated as excellent or good can be used for irrigation and generally will not cause soil salinization. For water rated as moderate, irrigation practices such as mode, amount, and crop type should be carefully considered.

The irrigation water quality of the main valley surface waters in the northern Qinling Mountains was evaluated based on the salt and alkali hazards indicated by individual factors. The Wilcox diagram and USSL diagram provide a

Table 5.5 Irrigation water quality evaluation results

ID	EC	SAR	RSC	%Na	PI	ID	EC	SAR	RSC	%Na	PI
QL1-01	118	0.19	−0.6	10.61	68.33	QL2-01	438	0.24	−2.21	7.75	38.67
QL1-02	147	0.12	−0.61	5.96	65.25	QL2-02	862	0.5	−5.72	11.23	27.43
QL1-03	516	0.7	−0.32	18.82	56.17	QL2-03	201	0.17	−0.91	7.84	56.3
QL1-04	471	0.68	−0.52	18.68	56.35	QL2-04	1143	3.33	−0.83	48.44	71.73
QL1-05	450	0.58	−0.43	17.02	57.21	QL2-05	1126	3.09	−1.33	45.86	68.21
QL1-06	501	0.98	−0.53	25.55	61.19	QL2-06	726	2.16	−1.23	41.12	65.12
QL1-07	710	2.19	−0.63	43.1	69.77	QL2-08	149	0.27	−0.3	14.15	82.01
QL1-08	376	0.2	−1.42	6.69	45.04	QL2-09	204	0.45	−0.5	19.82	73.37
QL1-09	860	1.52	−0.63	29.23	58.86	QL2-10	264	0.62	−0.41	22.06	69.66
QL1-10	655	2.01	−0.42	39.35	73.18	QL2-11	374	1.08	−0.31	32.14	73.25
QL1-11	614	2.14	−0.22	40.85	76.16	QL2-12	191	0.29	−0.41	13.1	71.35
QL1-12	1136	3.91	0.08	52.87	77.51	QL2-13	255	0.5	−0.27	19.39	67.12
QL1-13	656	2.08	−0.63	43.04	70.96	QL2-14	246	0.68	−0.23	27.05	65.18
QL1-14	555	3.65	0.63	64.11	92.21	QL2-15	363	0.4	−0.82	12.82	54.01
QL1-15	619	2.07	1.62	40.86	70.81	QL2-16	155	0.2	−0.21	10.11	80.29
QL1-16	694	2.51	1.47	46.12	76.28	QL2-17	193	0.19	−0.24	8.79	67.37
QL1-17	485	0.76	0.08	20.88	60.8	QL2-18	143	0.18	−0.33	9.75	71.85
QL1-18	165	0.16	−0.56	8.25	64.35	QL2-19	206	0.18	−0.02	8.45	72.19
QL1-19	116	0.17	−0.3	10.01	83.04	QL2-20	247	0.47	−0.24	17.9	67.89
QL1-20	134	0.2	−0.2	11.27	87.81	QL2-21	165	0.17	−0.31	8.27	71.86
QL1-21	160	0.21	−0.3	10.51	75.79	QL2-22	253	0.17	−0.31	7.04	62.98
QL1-22	777	2.07	−1.03	40.06	65.6	QL2-23	304	0.21	−0.72	7.64	54.2
QL1-23	214	0.22	−0.49	9.23	64.3	QL2-24	745	3.57	−0.3	58.56	82.59
QL1-24	179	0.2	−0.21	9.57	76.23	QL2-25	366	0.47	−0.41	15.16	59.58
QL1-25	201	0.19	−0.31	8.85	70.68	QL2-26	130	0.11	−0.2	6.09	79.5
QL1-26	363	0.21	−0.77	7.34	51.32	QL2-27	131	0.11	0	6.13	88.39
QL1-27	476	0.22	−0.65	8.74	59.5	QL2-28	117	0.1	−0.1	5.72	87.87
QL1-28	360	0.27	−0.61	9.21	54.37	QL2-29	132	0.1	0	5.81	91.69
QL1-29	349	0.34	−0.61	11.09	54.95	QL2-30	113	0.09	−0.1	5.28	87.82
QL1-30	564	1.36	−0.82	31.89	63.34	QL2-31	299	0.29	−0.66	10.48	57.26
QL1-31	363	0.5	−0.62	16.29	59.53	QL2-32	96	0.11	−0.14	6.92	96.97
QL1-32	115	0.09	−0.3	5.53	82.27	QL2-33	165	0.12	0	6.45	82.67
QL1-33	119	0.09	−0.2	5.5	84.02	QL2-34	223	0.14	−0.61	6.25	60.76
QL1-34	252	0.14	−0.81	5.52	54.25	QL2-36	135	0.11	−0.4	6.13	74.58
QL1-35	157	0.2	−0.1	10.12	83.27	QL2-37	300	0.21	−0.71	8.11	56.11
QL1-36	261	0.13	−0.91	5.38	53.5	QL2-38	505	0.54	−1.21	15.47	51.08
QL1-37	348	0.09	−0.81	3.04	48.01	QL2-39	164	0.15	−0.2	7.93	79.92
QL1-38	212	0.11	−0.3	4.99	68.1	QL2-41	372	0.07	−1.11	2.39	44.51
QL1-39	288	0.11	−0.81	4.27	52.46	QL2-42	208	0.11	−0.61	5.16	61.06
QL1-40	237	0.1	−0.61	4.42	58.47	QL2-43	204	0.1	−0.71	4.48	57.44
QL1-41	347	0.07	−0.97	2.59	46.8	QL2-44	129	0.09	−0.3	5.23	77.96
QL1-42	248	0.14	−0.57	5.93	59.7	QL2-45	233	0.11	−0.41	4.74	62.7
QL1-43	187	0.13	−0.29	5.98	70.11	QL2-46	221	0.17	−0.61	7.04	59.66

Table 5.5 (continued)

ID	EC	SAR	RSC	%Na	PI	ID	EC	SAR	RSC	%Na	PI
QL1-44	154	0.09	−0.31	4.63	73.98	QL2-47	134	0.09	−0.4	4.87	72.78
QL1-45	318	0.09	−0.61	3.26	52.5	QL2-48	145	0.11	−0.1	5.99	82.48
QL1-46	333	0.11	−0.59	4.04	52.46	QL2-49	444	0.11	−1.32	3.29	41.1
QL1-47	158	0.09	−0.31	4.64	72.45	QL2-50	230	0.08	−0.6	3.59	58.21
QL1-48	270	0.57	−0.51	20.52	67.71	QL2-51	168	0.2	−0.2	9.98	78.33
QL1-49	155	0.18	−0.2	9.78	84.9	QL2-52	200	0.24	−0.41	10.19	66.93
QL1-50	316	0.11	−1.21	4.15	46.67	QL2-53	976	1.4	−3.03	26.06	47.62
QL1-51	275	0.31	−0.41	11.84	64.8	QL2-54	216	0.74	−0.21	28.92	82.96
QL1-52	168	0.17	−0.5	8.36	69.42	QL2-55	552	1.84	−0.81	40.81	69.46
QL1-53	250	0.17	−0.91	6.81	54.23	QL2-56	223	0.32	−0.51	12.86	64.53
QL1-54	117	0.11	−0.3	6.57	80.34	QL2-57	225	0.37	−0.21	14.82	70.76
QL1-55	160	0.18	−0.3	8.74	77.15	QL2-58	206	0.33	−0.2	13.66	71.61
QL1-56	110	0.12	−0.3	7.11	82.53	QL2-59	152	0.15	−0.51	7.87	70.07
QL1-57	94	0.09	−0.3	5.69	84.51	QL2-60	180	0.2	−0.21	9.87	78.22
QL1-58	151	0.16	−0.7	8.45	62.96	QL2-61	152	0.19	−0.6	9.74	67.87
QL1-59	134	0.19	−0.2	10.29	82.62	QL2-62	296	0.35	−0.41	13.25	63.49
QL1-60	915	3.35	−1.61	54.44	72.68	QL2-63	393	0.45	−0.41	15.3	61.52
QL1-61	154	0.25	−0.3	12.39	78.02	QL2-64	283	0.33	−0.41	13.08	66.1
QL1-62	172	0.24	−0.4	11.63	73.39	QL2-65	184	0.31	−0.51	13.73	69.19
QL1-63	276	0.83	−0.21	29.38	79.09	QL2-66	473	0.61	−1.11	17.51	53.02
QL1-64	706	1.14	−1.82	25.74	53.13	QL2-67	218	0.46	−0.67	18.8	59.21
QL1-65	223	0.28	−0.51	12.18	67.72	QL2-68	332	0.46	−0.8	16.37	53.15
QL1-66	166	0.19	−0.4	9.59	74.19	QL2-69	286	0.56	−0.47	20.17	62.73
QL1-67	242	0.26	−0.57	10.89	57.57	QL2-70	250	0.3	−0.61	11.61	60.95
QL1-68	233	0.25	−0.61	11.17	65.14	QL2-71	267	0.46	−0.41	17.52	67.87
QL1-69	197	0.18	−0.61	8.92	65.64	QL2-72	166	0.16	−0.51	7.93	68.11
QL1-70	251	0.26	−0.71	10.54	60.27	QL2-73	172	0.21	−0.4	10.39	72.99
QL1-71	293	0.21	−0.41	8.18	60.4	QL2-74	554	0.52	−1.13	13.76	47.82
QL1-72	313	0.29	−0.72	10.56	56.19	QL2-75	347	0.78	−0.61	23.34	65.76
QL1-73	367	0.79	−0.32	24.18	68.75	QL2-76	228	0.28	−0.61	11.29	63
QL1-74	685	2.24	−0.72	43.91	71.56	QL2-77	770	2.38	−0.02	41.45	73.69

comprehensive assessment of both salt and alkali hazards (Figs. 5.2 and 5.3). The Wilcox diagram simultaneously reflects %Na and EC and is divided into five zones: Zone I (excellent irrigation water quality), Zone II (good water quality), Zone III (acceptable irrigation water quality), Zone IV (water quality to be retained), and Zone V (unsuitable irrigation water quality). As shown in the diagram, all water sample points fall between Zone I and Zone II; nine samples are in the good water quality zone, while the rest are in the excellent irrigation water quality zone. This indicates that the surface water in the study area poses no risk of salt or alkali hazards when used for agricultural irrigation. The USSL diagram shows that water samples are distributed in zones C1S1, C2S1, and C3S1. Samples in zones C1S1

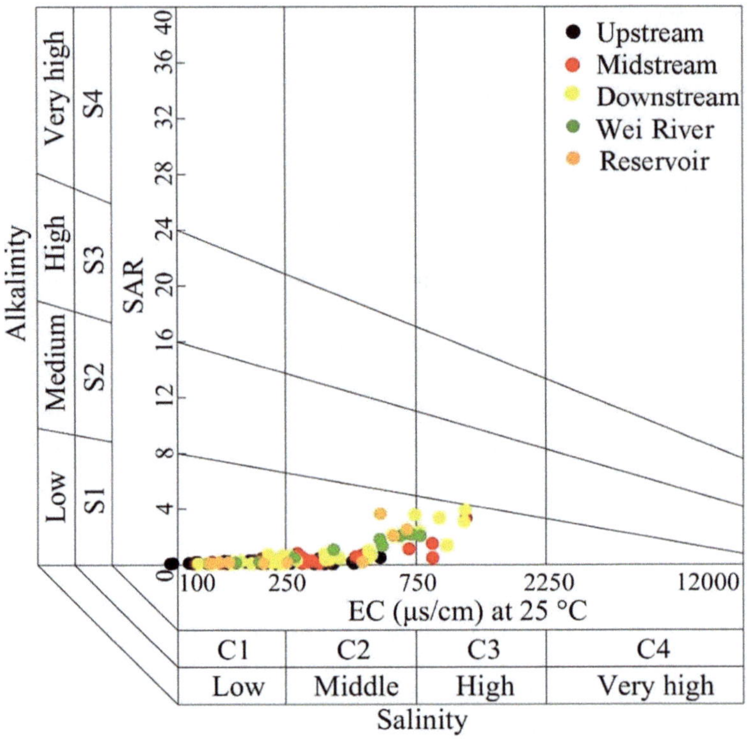

Fig. 5.2 Irrigation water quality USSL diagram

and C2S1 are suitable for irrigation, whereas those in C3S1 have slightly higher salt hazards and are less suitable for irrigation.

5.3.2.3 Industrial Water Quality Evaluation

In industrial water use, boiler water constitutes the primary part of the water supply, and its quality is governed by the standards set forth in the "*Water Quality Standard for Industrial Boilers of the People's Republic of China*" (GB/T 1576-2008). Due to the high temperature and high pressure conditions within steam boilers, certain substances in the water can undergo various adverse chemical reactions. Among these, scaling, foaming, and corrosion are detrimental chemical effects that seriously affect the normal operation of boilers. These issues not only lead to resource wastage and reduced boiler efficiency but also pose safety risks to boiler operation. Therefore, a water quality evaluation of boiler water was conducted. The evaluation methods are introduced as follows (Nsabimana et al. 2023):

1. Scaling potential

(a) Total Boiler Scale (H_0)

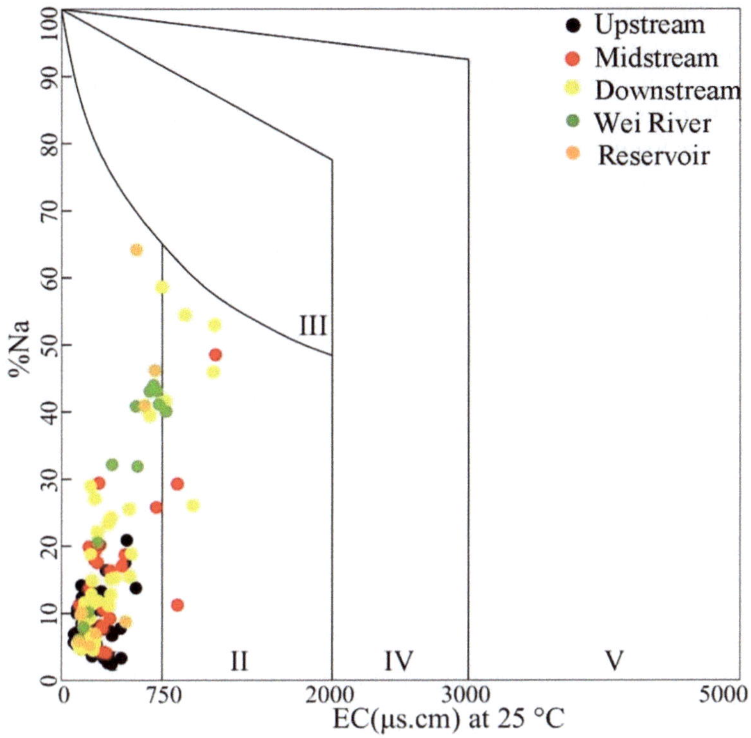

Fig. 5.3 Wilcox diagram for irrigation water quality

H_0 can be simply calculated using Eq. (5.5).

$$H_0 = S + C + 72\left[Fe^{2+}\right] + 51\left[Al^{3+}\right] + 40\left[Mg^{2+}\right] + 118\left[Ca^{2+}\right] \tag{5.5}$$

Where, H_0 is the total boiler scale content (mg/L); S is the suspended solids content in the water (mg/L); C is the colloid content (mg/L); the remaining terms represent ion concentrations (mmol/L).

(b)Total Hard Scale Content (H_h)

H_h can be computed using Eq. (5.6).

$$H_h = SiO_2 + 40\left[Mg^{2+}\right] + 68\left(\left[Cl^-\right] + 2\left[SO_4^{2-}\right] - \left[Na^+\right] - \left[K^+\right]\right) \tag{5.6}$$

(c)Hard Scale Coefficient (K_n)

K_n represents the ratio between H_h and H_0, and can be estimated using Eq. (5.7).

$$K_n = \frac{H_h}{H_0} \tag{5.7}$$

2. Foaming tendency

Due to the presence of certain special components in the water, a large amount of foaming occurs during boiling in the boiler, causing vaporization that leads to a rapid drop in the water level and results in the boiler's inability to operate normally. The foaming tendency can be evaluated using the foaming coefficient (*F*), expressed as Eq. (5.9).

$$F = 62\left[Na^{+}\right] + 78\left[K^{+}\right] \tag{5.8}$$

3. Corrosion effect

The corrosiveness of water can be evaluated using the corrosion coefficient (K_k). For acidic water and alkaline water, different equations are used (Eqs. 5.9 and 5.10).

(a) Acidic water

$$K_k = 1.008\left(\left[H^{+}\right] + 3\left[Al^{3+}\right] + 2\left[Fe^{2+}\right] + 2\left[Mg^{2+}\right] - \left[HCO_3^{-}\right] - 2\left[CO_3^{2-}\right]\right) \tag{5.9}$$

(b) Alkaline water

$$K_k = 1.008\left(2\left[Mg^{2+}\right] - \left[HCO_3^{-}\right]\right) \tag{5.10}$$

The water quality evaluation standards and results for industrial boiler water in the northern piedmont area of the Qinling Mountains are shown in Tables 5.6 and 5.7. According to the evaluation statistics, 110 water samples (70.97% of the total) are suitable for boiler use; 21 samples (13.55%) are basically suitable; and 24 samples (15.48%) are unsuitable for boiler use.

Table 5.6 Water quality evaluation standards for industrial boiler water

Total boiler scale H_0 (mg/L)	Hard scale coefficient (K_n)	Foaming tendency (F)	Corrosion coefficient (K_k)
$H_0 < 125$ mg/L Very low scale formation	$K_n < 0.25$ Soft precipitates	$F < 60$ mg/L No foaming	$K_k > 0$ Corrosive
125–250 mg/L Low scale formation	0.25–0.5 Medium precipitates	60–200 mg/L Partial foaming	$K_k < 0$ and $\Delta = K_k + 0.0503$ $Ca^{2+} > 0$ Moderately corrosive
250–500 mg/L Moderate scale formation	$K_n > 0.5$ Hard precipitates	$F > 200$ mg/L Foaming	$K_k < 0$ and $\Delta = K_k + 0.0503$ $Ca^{2+} < 0$ Non-corrosive
$H_0 > 500$ mg/L High scale formation			

Table 5.7 Water quality evaluation results for industrial boiler water

ID	Description	Suitability for boiler water use
QL1-01	Very low scale formation-Hard precipitates-No foaming-Non-corrosive	Not suitable
QL1-02	Very low scale formation-Medium precipitates-No foaming-Non-corrosive	Suitable
QL1-03	Low scale formation-Medium precipitates-Partial foaming-Non-corrosive	Suitable
QL1-04	Low scale formation-Medium precipitates-Partial foaming-Non-corrosive	Suitable
QL1-05	Low scale formation-Medium precipitates-No foaming-Non-corrosive	Suitable
QL1-06	Low scale formation-Medium precipitates-Partial foaming-Non-corrosive	Suitable
QL1-07	Low scale formation-Hard precipitates-Partial foaming-Non-corrosive	Not suitable
QL1-08	Low scale formation-Hard precipitates-No foaming-Non-corrosive	Not suitable
QL1-09	Moderate scale formation-Medium precipitates-Foaming-Non-corrosive	Basically suitable
QL1-10	Low scale formation-Medium precipitates-Foaming-Non-corrosive	Basically suitable
QL1-11	Low scale formation-Medium precipitates-Foaming-Non-corrosive	Basically suitable
QL1-12	Low scale formation-Soft precipitates-Foaming-Non-corrosive	Basically suitable
QL1-13	Low scale formation-Medium precipitates-Partial foaming-Non-corrosive	Suitable
QL1-14	Very low scale formation-Soft precipitates-Foaming-Non-corrosive	Basically suitable
QL1-15	Low scale formation-Soft precipitates-Partial foaming-Non-corrosive	Suitable
QL1-16	Very low scale formation-Soft precipitates-Foaming-Non-corrosive	Basically suitable
QL1-17	Low scale formation-Medium precipitates-Partial foaming-Non-corrosive	Suitable
QL1-18	Very low scale formation-Hard precipitates-No foaming-Non-corrosive	Not suitable
QL1-19	Very low scale formation-Hard precipitates-No foaming-Non-corrosive	Not suitable
QL1-20	Very low scale formation-Medium precipitates-No foaming-Non-corrosive	Suitable
QL1-21	Very low scale formation-Medium precipitates-No foaming-Non-corrosive	Suitable
QL1-22	Low scale formation-Hard precipitates-Foaming-Non-corrosive	Not suitable
QL1-23	Very low scale formation-Medium precipitates-No foaming-Non-corrosive	Suitable

(continued)

Table 5.7 (continued)

ID	Description	Suitability for boiler water use
QL1-24	Very low scale formation-Soft precipitates-No foaming-Non-corrosive	Suitable
QL1-25	Very low scale formation-Medium precipitates-No foaming-Non-corrosive	Suitable
QL1-26	Low scale formation-Medium precipitates-No foaming-Non-corrosive	Suitable
QL1-27	Low scale formation-Medium precipitates-No foaming-Non-corrosive	Suitable
QL1-28	Low scale formation-Medium precipitates-No foaming-Non-corrosive	Suitable
QL1-29	Low scale formation-Medium precipitates-No foaming-Non-corrosive	Suitable
QL1-30	Low scale formation-Medium precipitates-Partial foaming-Non-corrosive	Suitable
QL1-31	Low scale formation-Medium precipitates-No foaming-Non-corrosive	Suitable
QL1-32	Very low scale formation-Medium precipitates-No foaming-Non-corrosive	Suitable
QL1-33	Very low scale formation-Hard precipitates-No foaming-Non-corrosive	Not suitable
QL1-34	Low scale formation-Medium precipitates-No foaming-Non-corrosive	Suitable
QL1-35	Very low scale formation-Soft precipitates-No foaming-Non-corrosive	Suitable
QL1-36	Low scale formation-Medium precipitates-No foaming-Non-corrosive	Suitable
QL1-37	Low scale formation-Medium precipitates-No foaming-Non-corrosive	Suitable
QL1-38	Very low scale formation-Medium precipitates-No foaming-Non-corrosive	Suitable
QL1-39	Low scale formation-Medium precipitates-No foaming-Non-corrosive	Suitable
QL1-40	Low scale formation-Medium precipitates-No foaming-Non-corrosive	Suitable
QL1-41	Low scale formation-Medium precipitates-No foaming-Non-corrosive	Suitable
QL1-42	Low scale formation-Medium precipitates-No foaming-Non-corrosive	Suitable
QL1-43	Very low scale formation-Medium precipitates-No foaming-Non-corrosive	Suitable
QL1-44	Very low scale formation-Soft precipitates-No foaming-Non-corrosive	Suitable
QL1-45	Low scale formation-Medium precipitates-No foaming-Non-corrosive	Suitable
QL1-46	Low scale formation-Medium precipitates-No foaming-Non-corrosive	Suitable

(continued)

Table 5.7 (continued)

ID	Description	Suitability for boiler water use
QL1-47	Very low scale formation-Medium precipitates-No foaming-Non-corrosive	Suitable
QL1-48	Very low scale formation-Medium precipitates-No foaming-Non-corrosive	Suitable
QL1-49	Very low scale formation-Soft precipitates-No foaming-Non-corrosive	Suitable
QL1-50	Low scale formation-Hard precipitates-No foaming-Non-corrosive	Not suitable
QL1-51	Low scale formation-Medium precipitates-No foaming-Non-corrosive	Suitable
QL1-52	Very low scale formation-Medium precipitates-No foaming-Non-corrosive	Suitable
QL1-53	Low scale formation-Medium precipitates-No foaming-Non-corrosive	Suitable
QL1-54	Very low scale formation-Medium precipitates-No foaming-Non-corrosive	Suitable
QL1-55	Very low scale formation-Soft precipitates-No foaming-Non-corrosive	Suitable
QL1-56	Very low scale formation-Medium precipitates-No foaming-Non-corrosive	Suitable
QL1-57	Very low scale formation-Medium precipitates-No foaming-Non-corrosive	Suitable
QL1-58	Very low scale formation-Medium precipitates-No foaming-Non-corrosive	Suitable
QL1-59	Very low scale formation-Medium precipitates-No foaming-Non-corrosive	Suitable
QL1-60	Low scale formation-Hard precipitates-Foaming-Non-corrosive	Not suitable
QL1-61	Very low scale formation-Medium precipitates-No foaming-Non-corrosive	Suitable
QL1-62	Very low scale formation-Medium precipitates-No foaming-Non-corrosive	Suitable
QL1-63	Very low scale formation-Soft precipitates-No foaming-Non-corrosive	Suitable
QL1-64	Low scale formation-Medium precipitates-Partial foaming-Non-corrosive	Suitable
QL1-65	Very low scale formation-Medium precipitates-No foaming-Non-corrosive	Suitable
QL1-66	Very low scale formation-Medium precipitates-No foaming-Non-corrosive	Suitable
QL1-67	Very low scale formation-Medium precipitates-No foaming-Moderately corrosive	Basically suitable
QL1-68	Very low scale formation-Medium precipitates-No foaming-Non-corrosive	Suitable
QL1-69	Very low scale formation-Medium precipitates-No foaming-Non-corrosive	Suitable

(continued)

Table 5.7 (continued)

ID	Description	Suitability for boiler water use
QL1-70	Very low scale formation-Medium precipitates-No foaming-Non-corrosive	Suitable
QL1-71	Low scale formation-Soft precipitates-No foaming-Non-corrosive	Suitable
QL1-72	Low scale formation-Medium precipitates-No foaming-Non-corrosive	Suitable
QL1-73	Low scale formation-Medium precipitates-Partial foaming-Non-corrosive	Suitable
QL1-74	Low scale formation-Medium precipitates-Foaming-Non-corrosive	Basically suitable
QL2-01	Low scale formation-Hard precipitates-No foaming-Non-corrosive	Not suitable
QL2-02	Moderate scale formation-Hard precipitates-Partial foaming-Non-corrosive	Not suitable
QL2-03	Very low scale formation-Hard precipitates-No foaming-Non-corrosive	Not suitable
QL2-04	Moderate scale formation-Medium precipitates-Foaming-Non-corrosive	Basically suitable
QL2-05	Moderate scale formation-Medium precipitates-Foaming-Non-corrosive	Basically suitable
QL2-06	Low scale formation-Medium precipitates-Foaming-Non-corrosive	Basically suitable
QL2-08	Very low scale formation-Medium precipitates-No foaming-Non-corrosive	Suitable
QL2-09	Very low scale formation-Medium precipitates-No foaming-Non-corrosive	Suitable
QL2-10	Low scale formation-Medium precipitates-No foaming-Non-corrosive	Suitable
QL2-11	Low scale formation-Medium precipitates-Partial foaming-Non-corrosive	Suitable
QL2-12	Very low scale formation-Hard precipitates-No foaming-Non-corrosive	Not suitable
QL2-13	Very low scale formation-Medium precipitates-No foaming-Moderately corrosive	Basically suitable
QL2-14	Very low scale formation-Medium precipitates-No foaming-Moderately corrosive	Basically suitable
QL2-15	Low scale formation-Hard precipitates-No foaming-Non-corrosive	Not suitable
QL2-16	Very low scale formation-Medium precipitates-No foaming-Non-corrosive	Suitable
QL2-17	Very low scale formation-Medium precipitates-No foaming-Moderately corrosive	Basically suitable
QL2-18	Very low scale formation-Medium precipitates-No foaming-Moderately corrosive	Basically suitable

(continued)

Table 5.7 (continued)

ID	Description	Suitability for boiler water use
QL2-19	Very low scale formation-Soft precipitates-No foaming-Moderately corrosive	Basically suitable
QL2-20	Very low scale formation-Medium precipitates-No foaming-Moderately corrosive	Basically suitable
QL2-21	Very low scale formation-Medium precipitates-No foaming-Non-corrosive	Suitable
QL2-22	Low scale formation-Medium precipitates-No foaming-Non-corrosive	Suitable
QL2-23	Low scale formation-Medium precipitates-No foaming-Non-corrosive	Suitable
QL2-24	Low scale formation-Soft precipitates-Foaming-Non-corrosive	Suitable
QL2-25	Low scale formation-Medium precipitates-No foaming-Non-corrosive	Suitable
QL2-26	Very low scale formation-Medium precipitates-No foaming-Non-corrosive	Suitable
QL2-27	Very low scale formation-Soft precipitates-No foaming-Non-corrosive	Suitable
QL2-28	Very low scale formation-Medium precipitates-No foaming-Non-corrosive	Suitable
QL2-29	Very low scale formation-Soft precipitates-No foaming-Non-corrosive	Suitable
QL2-30	Very low scale formation-Medium precipitates-No foaming-Non-corrosive	Suitable
QL2-31	Low scale formation-Medium precipitates-No foaming-Non-corrosive	Suitable
QL2-32	Very low scale formation-Medium precipitates-No foaming-Non-corrosive	Suitable
QL2-33	Very low scale formation-Medium precipitates-No foaming-Non-corrosive	Suitable
QL2-34	Very low scale formation-Medium precipitates-No foaming-Non-corrosive	Suitable
QL2-36	Very low scale formation-Hard precipitates-No foaming-Non-corrosive	Not suitable
QL2-37	Low scale formation-Medium precipitates-No foaming-Non-corrosive	Suitable
QL2-38	Low scale formation-Medium precipitates-Partial foaming-Non-corrosive	Suitable
QL2-39	Very low scale formation-Medium precipitates-No foaming-Non-corrosive	Suitable
QL2-41	Low scale formation-Medium precipitates-No foaming-Non-corrosive	Suitable
QL2-42	Very low scale formation-Medium precipitates-No foaming-Non-corrosive	Suitable
QL2-43	Very low scale formation-Medium precipitates-No foaming-Non-corrosive	Suitable

(continued)

Table 5.7 (continued)

ID	Description	Suitability for boiler water use
QL2-44	Very low scale formation-Medium precipitates-No foaming-Non-corrosive	Suitable
QL2-45	Very low scale formation-Soft precipitates-No foaming-Non-corrosive	Suitable
QL2-46	Very low scale formation-Medium precipitates-No foaming-Non-corrosive	Suitable
QL2-47	Very low scale formation-Medium precipitates-No foaming-Non-corrosive	Suitable
QL2-48	Very low scale formation-Soft precipitates-No foaming-Non-corrosive	Suitable
QL2-49	Low scale formation-Medium precipitates-No foaming-Non-corrosive	Suitable
QL2-50	Very low scale formation-Medium precipitates-No foaming-Non-corrosive	Suitable
QL2-51	Very low scale formation-Medium precipitates-No foaming-Non-corrosive	Suitable
QL2-52	Very low scale formation-Soft precipitates-No foaming-Non-corrosive	Suitable
QL2-53	Moderate scale formation-Hard precipitates-Partial foaming-Non-corrosive	Not suitable
QL2-54	Very low scale formation-Soft precipitates-No foaming-Non-corrosive	Suitable
QL2-55	Low scale formation-Hard precipitates-Partial foaming-Non-corrosive	Not suitable
QL2-56	Very low scale formation-Medium precipitates-No foaming-Non-corrosive	Suitable
QL2-57	Very low scale formation-Soft precipitates-No foaming-Non-corrosive	Suitable
QL2-58	Very low scale formation-Soft precipitates-No foaming-Non-corrosive	Suitable
QL2-59	Very low scale formation-Medium precipitates-No foaming-Non-corrosive	Suitable
QL2-60	Very low scale formation-Soft precipitates-No foaming-Non-corrosive	Suitable
QL2-61	Very low scale formation-Medium precipitates-No foaming-Non-corrosive	Suitable
QL2-62	Low scale formation-Soft precipitates-No foaming-Non-corrosive	Suitable
QL2-63	Low scale formation-Medium precipitates-No foaming-Non-corrosive	Suitable
QL2-64	Very low scale formation-Soft precipitates-No foaming-Non-corrosive	Suitable
QL2-65	Very low scale formation-Medium precipitates-No foaming-Non-corrosive	Suitable
QL2-66	Low scale formation-Medium precipitates-Partial foaming-Non-corrosive	Suitable

(continued)

Table 5.7 (continued)

ID	Description	Suitability for boiler water use
QL2-67	Very low scale formation-Medium precipitates-No foaming-Moderately corrosive	Basically suitable
QL2-68	Low scale formation-Medium precipitates-No foaming-Moderately corrosive	Basically suitable
QL2-69	Very low scale formation-Medium precipitates-No foaming-Moderately corrosive	Basically suitable
QL2-70	Low scale formation-Medium precipitates-No foaming-Non-corrosive	Suitable
QL2-71	Very low scale formation-Soft precipitates-No foaming-Non-corrosive	Suitable
QL2-72	Very low scale formation-Medium precipitates-No foaming-Non-corrosive	Suitable
QL2-73	Very low scale formation-Soft precipitates-No foaming-Non-corrosive	Suitable
QL2-74	Moderate scale formation-Medium precipitates-Partial foaming-Non-corrosive	Suitable
QL2-75	Low scale formation-Medium precipitates-Partial foaming-Non-corrosive	Suitable
QL2-76	Very low scale formation-Medium precipitates-No foaming-Non-corrosive	Suitable
QL2-77	Low scale formation-Soft precipitates-Foaming-Non-corrosive	Basically suitable
QL2-78	Low scale formation-Hard precipitates-No foaming-Non-corrosive	Not suitable
QL2-79	Moderate scale formation-Hard precipitates-Foaming-Non-corrosive	Not suitable
QL2-80	Low scale formation-Hard precipitates-Foaming-Non-corrosive	Not suitable
QL2-81	Moderate scale formation-Medium precipitates-Foaming-Non-corrosive	Not suitable
QL2-82	Moderate scale formation-Hard precipitates-Foaming-Non-corrosive	Not suitable
QL2-83	Low scale formation-Hard precipitates-No foaming-Non-corrosive	Not suitable
QL2-84	Low scale formation-Hard precipitates-Foaming-Non-corrosive	Not suitable

5.3.3 *Spatial Distribution Characteristics of Surface Water Quality*

5.3.3.1 Spatial Distribution of Drinking Water Quality

The spatial distribution of drinking water quality in the study area's surface water is shown in Fig. 5.4. Multiple rivers within the study area exhibit a spatial pattern of water quality deterioration from upstream to midstream and downstream. The

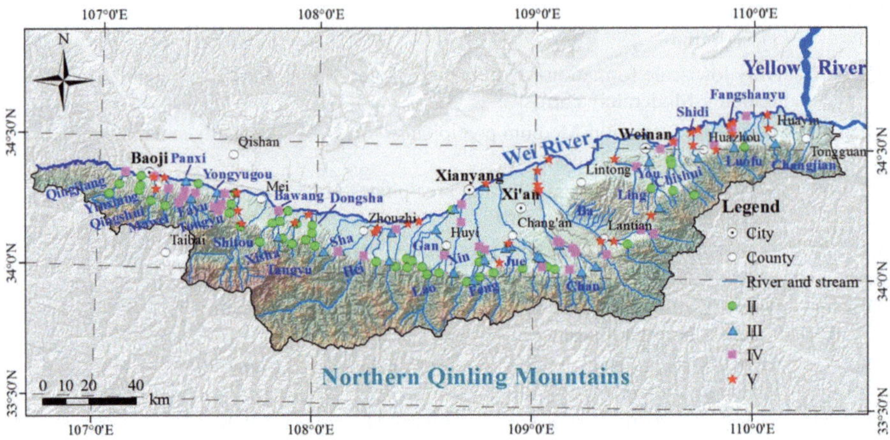

Fig. 5.4 Spatial distribution of drinking water quality

overall water quality of the Qingjiang River, Yinxiang River, Tangyu River, Bawang River, and Dagengyu River from upstream to downstream is classified as Grade II, indicating relatively good water quality suitable for drinking. In contrast, Shitou River and Hei River show overall water quality classified as Grade V from upstream to downstream, indicating poor water quality unsuitable for drinking. The water quality in the Wei River section is mostly Grades IV and V, reflecting poor water quality. This is likely due to the large drainage area of the Wei River basin, which passes through multiple cities and counties, with many tributaries flowing into the Wei River, carrying various impurities.

Baoji City, located in the upper reaches of the Wei River, has good water quality in multiple rivers within its jurisdiction. In contrast, Weinan City, situated in the lower reaches of the Wei River, exhibits poorer water quality compared to Baoji. Grade IV water samples account for 11% of the total samples in Baoji, while Grade V samples make up 19%. In Xi'an City, water quality is generally good, with Grade IV samples constituting 25% and Grade V samples 18% of the total. Weinan City shows worse water quality than both Baoji and Xi'an, with Grade IV samples accounting for 20% and Grade V samples 30% of its total water samples.

The water quality near the expressways and the Guanzhong Ring Road is mostly classified as Grade IV and V. This is because, upon exiting the Qinling piedmont and entering the Guanzhong Plain, the number of nearby villages, factories, and farmlands increases. In addition, the heavy traffic around the expressways and ring road generates road runoff that flows into surface waters, significantly increasing the total dissolved solids. This runoff also raises the concentrations of heavy metals and organic pollutants in the surface water, adversely affecting its quality.

5.3.3.2 Spatial Distribution of Irrigation Water Quality

The spatial distribution of irrigation water quality in the study area is shown in Fig. 5.5. Multiple rivers in the region exhibit a spatial trend of water quality gradually changing from suitable for irrigation in the upper reaches to basically suitable for irrigation in the middle and lower reaches. In Baoji City, located in the upper reaches of the Wei River, nearly all rivers are basically suitable for irrigation. In Xi'an City, water samples unsuitable for irrigation are mainly distributed downstream of the Xin River, Feng River, and Ba River; these three rivers are close to residential areas or industrial parks, where water quality is affected and thus unsuitable for irrigation. In Weinan City, which lies in the lower reaches of the Wei River, the number of water samples unsuitable for irrigation increases, including sites in the midstream Fangshan River, downstream You River (at the confluence with the Wei River), downstream Qingjiang River (at the confluence with the Wei River), midstream and downstream Changjian River, and midstream Luofu River.

Water quality in Baoji City is relatively good, with water samples suitable for irrigation accounting for 66% of the total samples in Baoji, and those basically suitable for irrigation accounting for 34%. In Xi'an City, water samples suitable for irrigation account for 58% of the total samples, basically suitable samples account for 38%, and those unsuitable for irrigation account for 4%. In Weinan City, water samples suitable for irrigation account for 58% of the total samples, basically suitable samples account for 45%, and samples unsuitable for irrigation account for 13%.

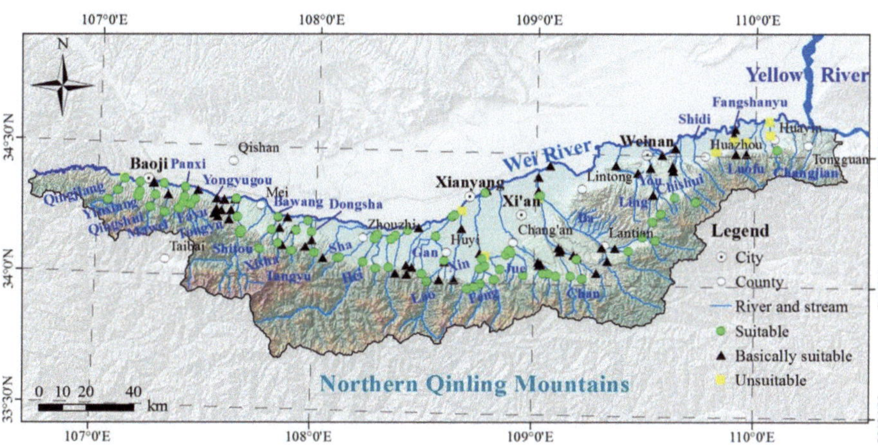

Fig. 5.5 Spatial distribution of irrigation water quality

5.3.3.3 Spatial Distribution Characteristics of Industrial Water Quality

The spatial distribution of industrial water quality in the study area is shown in Fig. 5.6. The figure indicates that industrial boiler water quality in the central and western parts of the region is generally good and suitable for industrial boiler use. Only a few rivers have water quality unsuitable for boiler use, including the upper reaches of the Qingjiang River, Mawei River, lower reaches of Tangyu River, upper reaches of the Xisha River, lower reaches of Xin River (near the confluence with the Wei River), lower reaches of the Jue River, and upper reaches of the Liyu River (a tributary of Lao River). In contrast, the water quality in the eastern part of the study area, within Weinan City, is generally poor, with more than half of the samples deemed unsuitable for industrial boiler use. Unsuitable samples are distributed across nearly all rivers, with only the You River Reservoir, Sanzhang Reservoir, and certain sections of the Fangshan River, You River, Chishui River, Changjian River, and Jianyu River showing water quality that is suitable or basically suitable for boiler use.

The majority of water samples within Baoji City are suitable for industrial use. Specifically, samples suitable for industrial boiler water account for 80% of the total samples in Baoji, while those basically suitable account for 6%, and those unsuitable make up 9%. In Xi'an City, 87% of the water samples are suitable for industrial use, 5% are basically suitable, and 4% are unsuitable. In contrast, the water quality in Weinan City is relatively poorer, with only 32% of samples suitable for industrial use, 23% basically suitable, and 14% unsuitable.

Fig. 5.6 Spatial distribution of industrial boiler water quality

5.4 Factors Influencing Surface Water Environment

There are many factors that influence changes in surface water environments, but changes in natural conditions and human activities are the primary driving forces. All influencing factors can be categorized into these two major groups. During this scientific investigation, it was found that the surface water bodies along the northern slope of the Qinling Mountains are affected by both natural processes and human activities. These two influences are intertwined, jointly controlling the changes in the surface water environment of the region.

5.4.1 Gibbs Diagram

The Gibbs diagram can visually reflect the formation mechanisms of major ions in natural water bodies. By analyzing the relationships between $Na^+/(Na^+ + Ca^{2+})$ and $Cl^-/(Cl^- + HCO_3^-)$ with TDS, the Gibbs diagram categorizes the dominant mechanisms controlling the major chemical components of natural waters into three types: evaporation dominance, precipitation dominance, and rock–water interaction dominance. The Gibbs diagram consists of two plots: one representing the relationship between $Na^+/(Na^+ + Ca^{2+})$ and TDS, and the other showing the relationship between $Cl^-/(Cl^- + HCO_3^-)$ and TDS. The Gibbs diagrams for the 156 surface water samples analyzed in this study are shown in Fig. 5.7.

The distribution of surface water samples in Fig. 5.7 shows that all samples fall within the middle portion of the Gibbs diagrams, indicating that the hydrochemical composition of surface water in the study area is primarily controlled by rock–water interaction. This interaction is a long-term and continuous process. The upward extension trend of sample points suggests that evaporation also plays a secondary role in shaping the hydrochemical characteristics of surface water in the region. Moreover, most sample points are located on the left side of the $Cl^-/(Cl^- + HCO_3^-)$ diagram, indicating that bicarbonate (HCO_3^-) is more abundant than chloride (Cl^-) in the surface water, and this finding is consistent with the results from some of the Piper diagram analyses.

To reveal the hydrochemical formation mechanisms of surface water in different areas, the water samples were divided into three regional groups from west to east: Baoji, Xi'an, and Weinan, as shown in Fig. 5.8. The red circles, blue diamonds, and green triangles represent surface water samples from the Baoji, Xi'an, and Weinan sections, respectively. Compared with the samples from Baoji and Xi'an, the Weinan section shows a more pronounced upward extension trend, indicating that the surface water in the Weinan section is more strongly influenced by evaporation in its hydrochemical formation than in the Baoji and Xi'an sections.

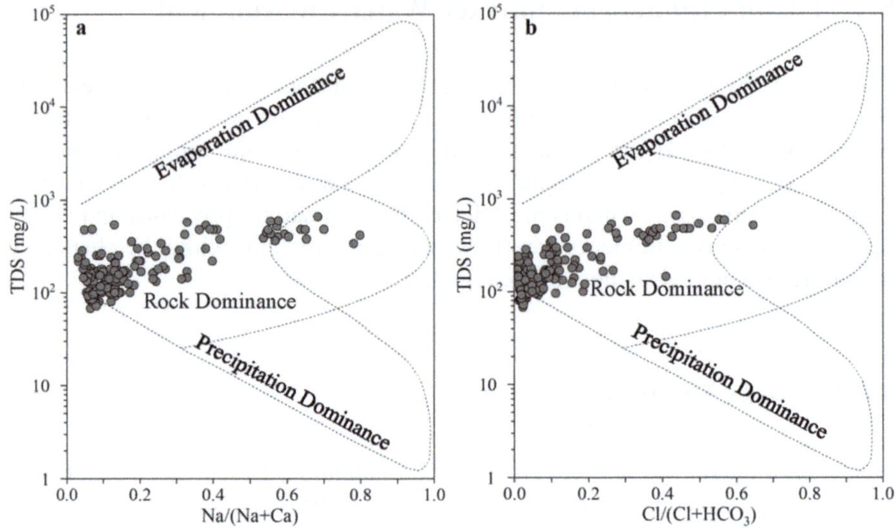

Fig. 5.7 Gibbs diagrams showing hydrochemical formation mechanisms of surface water

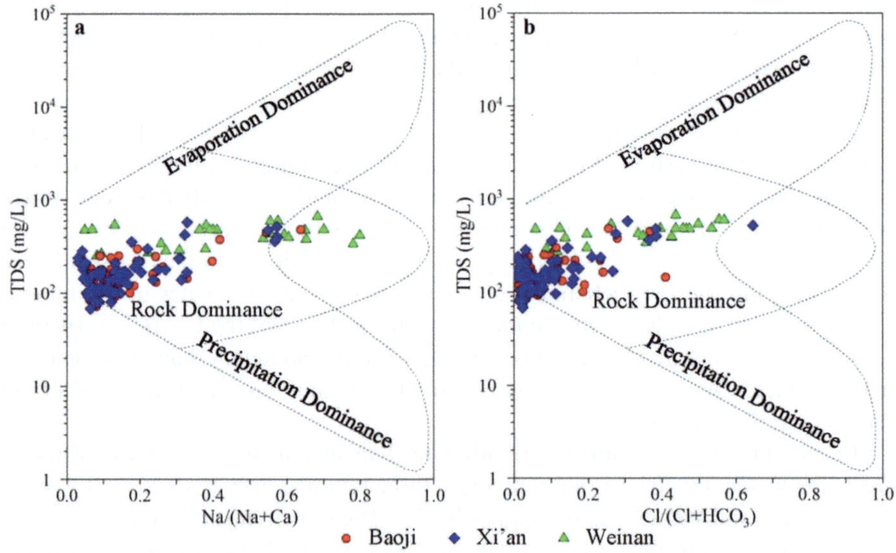

Fig. 5.8 Gibbs diagrams of water samples from different regions in the study area

To investigate whether there are differences in the hydrochemical formation of surface water from the upstream, midstream, and downstream sections of tributaries entering the Wei River, the water samples were categorized into five groups: "Upstream", "Midstream", "Downstream", "Wei River", and "Reservoir". The "Upstream", "Midstream" and "Downstream" groups represent surface water samples collected from different sections of the tributaries flowing into the Wei River, "Wei River" represents samples from the Wei River, and "Reservoir" represents

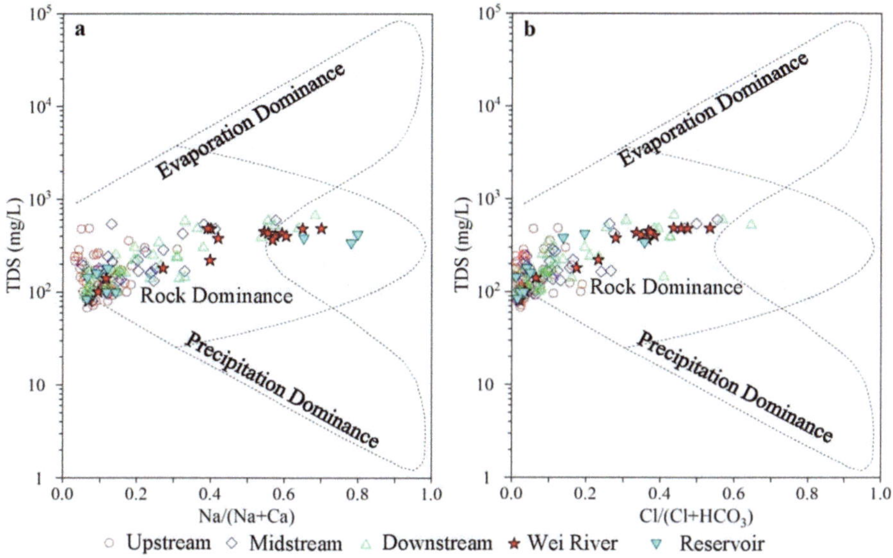

Fig. 5.9 Gibbs diagrams of river upstream, midstream, downstream, and water samples from the Wei River and reservoirs

samples collected from reservoirs within the study area. The Gibbs diagrams for these five categories are shown in Fig. 5.9. In the figure, red hollow circles, blue hollow diamonds, and green hollow upward triangles represent the upstream, midstream, and downstream tributary water samples, respectively. Red pentagrams represent Wei River water samples, and blue downward triangles represent reservoir water samples. The upstream tributary samples (red hollow circles) exhibit a weaker upward extension trend compared to other sample groups, indicating that the hydrochemical formation of upstream tributary surface waters in the study area is less influenced by evaporation.

Figure 5.10 shows the Gibbs diagrams of water samples collected before and after rainy season. In the figure, red circles represent water samples taken before rainfall, and blue diamond shapes represent samples taken after rainfall. Compared to post-rainfall samples, pre-rainfall samples exhibit a greater upward extension trend, indicating that the water chemistry of pre-rainfall samples is more influenced by evaporation. In addition, post-rainfall samples are more concentrated on the left side of the diagram, indicating that the concentrations of Na^+ and Cl^- in post-rainfall samples are lower than those in pre-rainy season samples.

5.4.2 *Correlation Analysis*

The Pearson correlation coefficient matrix is a very useful tool in hydrogeochemical studies, as it can reflect the relationships between each ion or parameter, thereby further characterizing the coexistence within the entire dataset or the connections

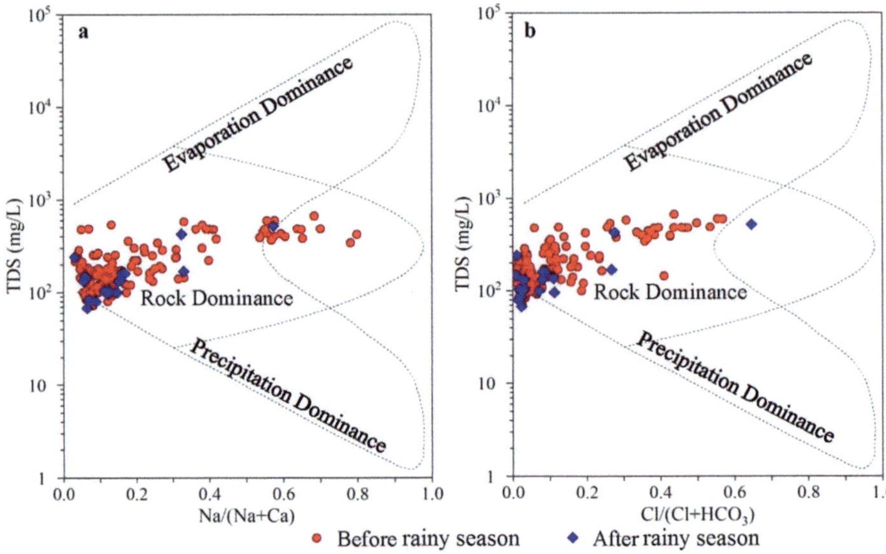

Fig. 5.10 Gibbs diagram of water samples before and after the rainy season

between individual indicators. In this study, SPSS was used to calculate the Pearson correlation coefficient matrix of various water quality indicators, as shown in Table 5.8.

According to Table 5.8, Cl^- shows a significant positive correlation with Na^+, and SO_4^{2-} also exhibits a significant positive correlation with Na^+, indicating that they share common sources or formation processes. Considering that the samples are from surface water, the weathering and dissolution of rock salt during surface water flow are the main sources of C^{l-} and Na^+ in the surface water. Additionally, the dissolution of mirabilite also affects the concentrations of SO_4^{2-} and Na^+ in the surface water. The related reactions are expressed as Eqs. (5.11) and (5.12):

$$NaCl \rightarrow Na^+ + Cl^- \tag{5.11}$$

$$Na_2SO_4 \rightarrow 2Na^+ + SO_4^{2-} \tag{5.12}$$

Ca^{2+} shows a significant positive correlation with SO_4^{2-}, indicating that they share common sources or formation processes. Therefore, the weathering and dissolution of gypsum ($CaSO_4{\cdot}2H_2O$) is likely one of the sources of Ca^{2+} and SO_4^{2-} in the surface water, as shown in the following reaction (Eq. 5.13):

$$CaSO_4{\cdot}2H_2O \rightarrow Ca^{2+} + SO_4^{2-} + 2H_2O \tag{5.13}$$

Table 5.8 Pearson correlation coefficient matrix of major surface water quality indicators

	pH	K^+	Na^+	Ca^{2+}	Mg^{2+}	F^-	Cl^-	SO_4^{2-}	HCO_3^-	CO_3^{2-}	NO_3^-	NO^{2-}	COD_{cr}	SiO_2	TDS	TH
pH	1.00	**−0.33**	**−0.29**	**−0.25**	−0.05	0.14	**−0.35**	**−0.25**	*−0.19*	**0.50**	*−0.19*	**−0.36**	**−0.25**	−0.04	**−0.29**	*−0.21*
K^+		1.00	**0.70**	**0.42**	**0.39**	*0.19*	**0.77**	**0.43**	**0.54**	−0.05	0.07	**0.55**	**0.64**	**0.35**	**0.69**	**0.47**
Na^+			1.00	**0.44**	**0.66**	**0.41**	**0.92**	**0.67**	**0.69**	*0.18*	0.10	**0.51**	**0.69**	**0.29**	**0.89**	**0.59**
Ca^{2+}				1.00	**0.50**	0.14	**0.54**	**0.74**	**0.69**	−0.11	**0.36**	**0.37**	**0.21**	**0.41**	**0.77**	**0.93**
Mg^{2+}					1.00	**0.40**	**0.53**	**0.56**	**0.87**	**0.30**	**0.22**	*0.17*	**0.45**	**0.27**	**0.77**	**0.78**
F^-						1.00	**0.31**	**0.29**	**0.32**	**0.30**	0.01	0.07	**0.25**	**0.32**	**0.38**	**0.27**
Cl^-							1.00	**0.61**	**0.60**	0.02	0.09	**0.47**	**0.63**	**0.35**	**0.85**	**0.61**
SO_4^{2-}								1.00	**0.55**	−0.06	**0.30**	**0.43**	**0.41**	**0.29**	**0.84**	**0.77**
HCO_3^-									1.00	0.12	*0.18*	**0.38**	**0.44**	**0.41**	**0.85**	**0.87**
CO_3^{2-}										1.00	*−0.17*	−0.07	*0.19*	−0.09	0.08	0.04
NO_3^-											1.00	0.01	0.08	0.01	**0.28**	**0.35**
NO_2^-												1.00	**0.46**	**0.24**	**0.51**	**0.34**
COD_{cr}													1.00	0.09	**0.60**	**0.35**
SiO_2														1.00	**0.42**	**0.41**
TDS															1.00	**0.88**
TH																1.00

Bold font indicates significance at the $\alpha = 0.01$ level; italics indicate significance at the $\alpha = 0.05$ level

COD_{Cr} in surface water shows varying degrees of positive correlations with K^+, Na^+, Ca^{2+}, Mg^{2+}, F^-, Cl^-, SO_4^{2-}, HCO_3^-, CO_3^{2-}, and NO_2^-. Additionally, nitrate (NO_3^-) exhibits positive correlations with Ca^{2+}, Mg^{2+}, and SO_4^{2-}. These correlations indicate that human activities have a direct or indirect impact on the concentrations of ions in the surface water.

The pH shows a significant positive correlation with CO_3^{2-}, indicating that alkaline conditions are favorable for the formation of carbonate ions. HCO_3^- is significantly positively correlated with Ca^{2+} and Mg^{2+}, suggesting that the formation of HCO_3^-, Ca^{2+}, and Mg^{2+} ions in surface water is closely related to the dissolution of minerals such as calcite and dolomite, as shown in the following equations (Eqs. 5.14 and 5.15):

$$CaCO_3 + CO_2 + H_2O \rightarrow Ca^{2+} + 2HCO_3^- \tag{5.14}$$

$$CaMg\left(CO_3\right)_2 + 2CO_2 + 2H_2O \rightarrow Mg^+ + Ca^{2+} + 4HCO_3^- \tag{5.15}$$

The scatter plot of the ion ratio relationships in water can be used to identify hydrogeochemical processes and formation mechanisms, as shown in Fig. 5.11. From Fig. 5.11a, most of the Na^+ and Cl^- data points lie above the 1:1 line,

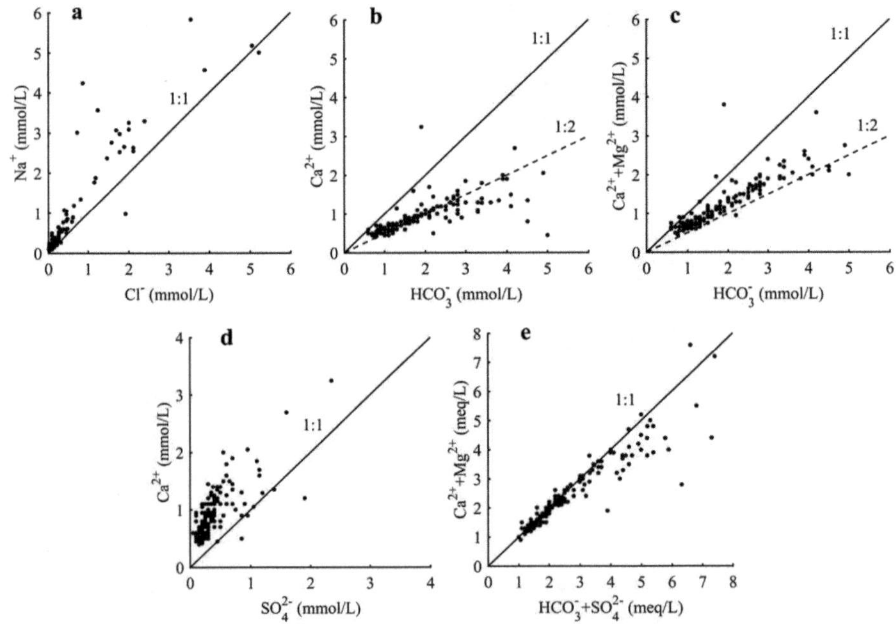

Fig. 5.11 Ion ratio relationships

indicating that the Na^+ concentration is higher than Cl^-. This suggests that the dissolution of halite is not the sole source of Na^+ and Cl^- in the river water; the excess Na^+ may also originate from the weathering and dissolution of crustal silicate minerals, such as sodium feldspar and sodium montmorillonite. As shown in Table 5.8, SiO_2 is positively correlated with K^+, Na^+, Ca^{2+}, Mg^{2+}, and HCO_3^-, indicating that silicate dissolution influences the concentrations of these ions. Moreover, during surface water flow, cation exchange processes may occur, affecting the concentrations of Na^+ and Ca^{2+}.

$$\begin{aligned}&4NaAlSi_3O_8 + 4CO_2 + 22H_2O \rightarrow Al_4Si_4O_{10}(OH)_8\\&+8H_4SiO_4 + 4Na^+ + 4HCO_3^-\end{aligned} \tag{5.16}$$

$$3Na_{1/3}Al_{7/3}Si_{11/3}O_{10} + 30H_2O + 6OH^- \rightarrow Na^+ + 7Al(OH)_4^- + 11H_4SiO_4 \tag{5.17}$$

$$H_4SiO_4 \rightarrow SiO_2 + 2H_2O \tag{5.18}$$

Ca^{2+}, Mg^{2+}, and HCO_3^- in surface water generally originate from the dissolution of carbonates such as calcite and dolomite. When the reaction occurs in an open system, CO_2 participates, causing the molar ratio of Ca^{2+} vs HCO_3^- to vary between 1:1 and 1:2. As shown in Fig. 5.11b, most water samples fall between the 1:1 and 1:2 lines, indicating that calcite dissolution is one of the sources of Ca^{2+} and HCO_3^- in surface water. However, a small portion of samples lie below the 1:2 line, suggesting that during river flow, Ca^{2+} may decrease due to cation exchange, while Na^+ content increases, further supporting the possibility of cation exchange. The dissolution of dolomite also varies with the amount of CO_2 involved in the reaction, resulting in the molar ratio of Ca^{2+} + Mg^{2+} vs HCO_3^- ranging between 1:1 and 1:2. Figure 5.11c shows that most samples lie between the 1:1 and 1:2 lines, indicating that dolomite dissolution is another source of Ca^{2+}, Mg^{2+}, and HCO_3^- in surface water. A few samples lie above the 1:1 line, suggesting that Ca^{2+} and Mg^{2+} also have other sources, such as gypsum dissolution. The scatter plot of Ca^{2+} versus SO_4^{2-} is shown in Fig. 5.11d, where most samples deviate significantly from the 1:1 line and lie above it, indicating that gypsum dissolution has a limited influence on Ca^{2+} and SO_4^{2-} concentrations in surface water. If calcite, dolomite, and gypsum are the main sources of Ca^{2+}, Mg^{2+}, HCO_3^-, and SO_4^{2-} in surface water, the milliequivalent concentration ratio of Ca^{2+}+Mg^{2+} vs HCO_3^- + SO_4^{2-} should be close to 1:1. Figure 5.11e shows that Ca^{2+}+Mg^{2+} and HCO_3^- + SO_4^{2-} mostly vary along the 1:1 line, but a small number of samples fall below it, indicating that the main source of Ca^{2+} and Mg^{2+} in surface water is the dissolution of calcite or dolomite, although cation exchange may reduce Ca^{2+} concentration and human activities may increase SO_4^{2-} content.

5.4.3 Multivariate Statistical Analysis

Multivariate statistical analysis is a method that uses mathematical statistics to handle problems involving multiple variables and has been widely applied in geology, hydrology, meteorology, medicine, water quality evaluation, hydrochemical composition analysis, and pollution source identification. Commonly used techniques include cluster analysis, factor analysis, principal component analysis, and discriminant analysis (Wu et al. 2019). In this study, cluster analysis was applied to investigate the surface water quality and hydrochemical composition in the study area.

Cluster analysis groups water quality indicators into a limited number of clusters based on their similarity, thereby exploring possible sources or influencing factors of different indicators. Initially, the raw data were transformed using Z-scores (standard normal transformation). Then, the chi-square distance metric was used to calculate the distance coefficient because it offers stronger discriminative ability compared to commonly used metrics like Euclidean distance. Finally, Ward's clustering method was applied for clustering. The entire process was conducted using SPSS. The cluster dendrogram is shown in Fig. 5.12.

As shown in Fig. 5.12, the main chemical indicators of surface water samples in the study area can be grouped into two categories. The first category (Category I)

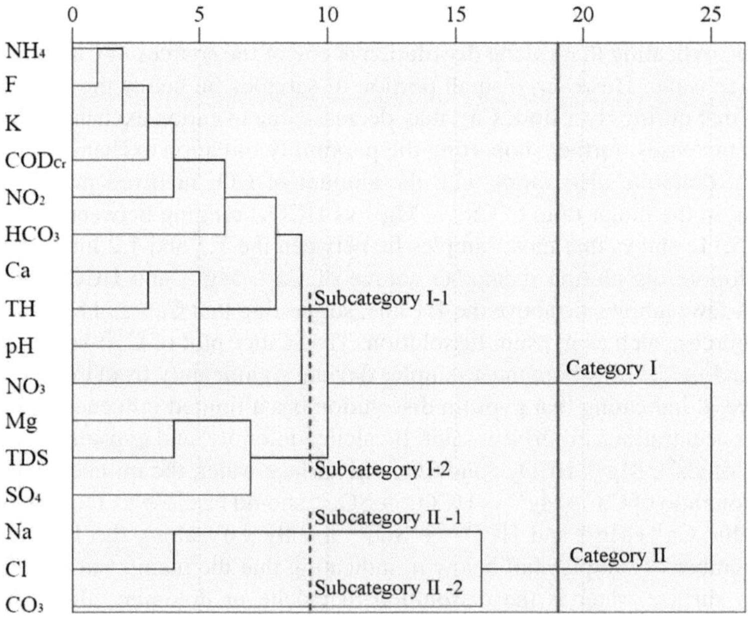

Fig. 5.12 Cluster dendrogram of major chemical indicators

includes NH_4^+, F^-, K^+, COD_{Cr}, NO_2^-, HCO_3^-, Ca^{2+}, TH, pH, NO_3^-, Mg^{2+}, TDS, and SO_4^{2-}. Category I is further divided into two subcategories: Subcategory I-1 and Subcategory I-2. Subcategory I-1 includes NH_4^+, F^-, K^+, COD_{Cr}, NO_2^-, HCO_3^-, Ca^{2+}, TH, pH, and NO_3^-. Among these, NH_4^+, NO_2^-, and NO_3^- mainly originate from human agricultural activities such as the use of fertilizers and pesticides. COD_{Cr} generally comes from domestic sewage and industrial wastewater. Alkaline environments favor the precipitation of CaF_2. High HCO_3^- indicates the dissolution of carbonates like calcite and dolomite. The dissolution of calcite and dolomite leads to increased concentrations of Ca^{2+} and Mg^{2+}. pH, TH, and K^+ usually originate from natural evolution processes of the surface water body. Therefore, Subcategory I-1 can be summarized as a combined influence of natural processes and human activities. Subcategory I-2 includes Mg^{2+}, TDS, and SO_4^{2-}. Generally, Mg^{2+}, TDS, and SO_4^{2-} come from natural evolution of surface water but can also be influenced by human activities. Hence, Subcategory I-2 also represents a combined influence of natural processes and human activities. Thus, Category I indicates the combined effects of natural processes and human activities. The second category (Category II) includes Na^+, Cl^-, and CO_3^{2-}. Category II is divided into two subcategories: Subcategory II-1 (Na^+ and Cl^-) and Subcategory II-2 (CO_3^{2-}). Typically, Na^+, Cl^-, and CO_3^{2-} originate from the natural evolution of surface water. Therefore, Category II can be summarized as representing the natural evolution of surface water. In summary, the surface water chemistry in the study area is influenced by both natural processes and human activities.

5.4.4 *Stable Isotope Analysis*

Isotope techniques have been widely applied in studies related to the water cycle. They can be used to identify the sources of surface water and groundwater, as well as exchanges between different water bodies (Zhang et al. 2022). In this study, a total of 151 hydrogen and oxygen stable isotope samples were analyzed, including three rainfall samples and 148 surface water samples.

The isotope statistics for different sections within the study area (Baoji, Xi'an, and Weinan) are shown in Table 5.9. According to Table 5.9, the $\delta^{18}O$ values of surface water samples in the Baoji section range from −10.15 to −6.70‰, with an average of −8.39‰; δD values range from −72.78 to −42.98‰, with an average of −53.62‰. In the Xi'an section, $\delta^{18}O$ and δD values range from −10.67 to −7.86‰ and −72.94 to −52.45‰, respectively, with corresponding averages of −9.20‰ and −60.49‰. For the Weinan section, $\delta^{18}O$ and δD values range from −9.71 to −0.89‰ and −66.85 to −20.51‰, respectively, with averages of −7.82‰ and −55.72‰. The overall consistency among these values indicates that the surface water in the study area is recharged by atmospheric precipitation. The hydrogen and oxygen

Table 5.9 Statistical results of isotopes in different areas (‰)

	Baoji		Xi'an		Weinan	
Statistics	$\delta^{18}O$	δD	$\delta^{18}O$	δD	$\delta^{18}O$	δD
Max	−6.70	−42.98	−7.86	−52.45	−0.89	−20.51
Min	−10.15	−72.78	−10.67	−72.94	−9.71	−66.85
Mean	−8.39	−53.62	−9.20	−60.49	−7.82	−55.72
Variance	0.67	4.91	0.53	3.88	1.75	8.87

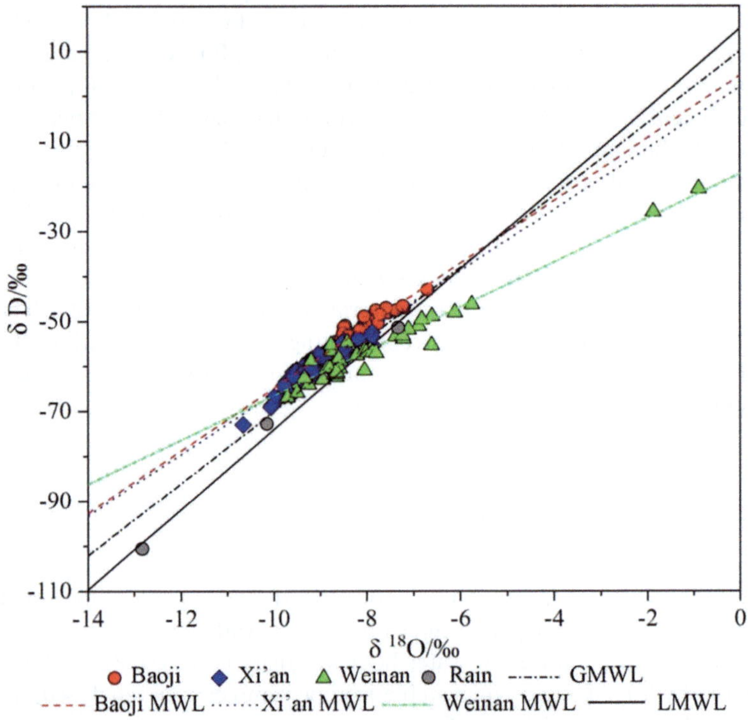

Fig. 5.13 Hydrogen and oxygen isotope relationship diagram of surface water

stable isotope relationships for each section of surface water in the study area are shown in Fig. 5.13. The equations of the rainfall line and the fitted lines for each water sample in Fig. 5.13 are as follows (Eqs. 5.19, 5.20, 5.21, 5.22 and 5.23):

$$\text{GMWL}: \delta D = 8.0 \times \delta^{18}O + 10.0 \tag{5.19}$$

$$\text{LMWL}: \delta D = 8.9043 \times \delta^{18}O + 15.077 \tag{5.20}$$

$$\text{Weinan MWL}: \delta D = 6.9435 \times \delta^{18}O + 4.6079 \quad (5.21)$$

$$\text{Xi'an MWL}: \delta D = 6.798 \times \delta^{18}O + 2.0276 \quad (5.22)$$

$$\text{Baoji MWL}: \delta D = 4.928 \times \delta^{18}O - 17.167 \quad (5.23)$$

Figure 5.13 shows that the local meteoric water line (LMWL) in the study area deviates from the global meteoric water line (GMWL), indicating that secondary evaporation occurs during precipitation in the region. Furthermore, the fitted lines for the water samples from each section deviate from the local meteoric water line, suggesting that evaporation continues during the recharge of surface water by precipitation (Pang et al. 2017). The $\delta^{18}O$ and δD compositions of surface water in the Baoji and Xi'an sections deviate less from the local meteoric water line, whereas those in the Weinan section show greater deviation. This implies that surface water in the Weinan section is more strongly affected by evaporation, consistent with the conclusions drawn from the Gibbs diagram.

To explore whether there are differences in the isotopic composition of surface water at the upstream, midstream, and downstream sections of tributaries flowing into the Wei River at various valleys, the water samples were also divided into five categories: "Upstream", "Midstream", "Downstream", "Wei River", and "Reservoir". The statistics for the hydrogen and oxygen stable isotopes of these five categories are shown in Table 5.10. According to Table 5.10, the $\delta^{18}O$ values of surface water from the upstream, midstream, and downstream sections of tributaries range from −10.00 to −7.72‰, −10.05 to −5.75‰, and −10.67 to −6.61‰, with averages of −8.91‰, −8.45‰, and −8.74‰, respectively; the δD values range from −66.84 to −48.50‰, −68.60 to −46.17‰, and −72.94 to −47.00‰, with averages of −57.90‰, −56.38‰, and −58.22‰, respectively. The $\delta^{18}O$ and δD values of the Wei River samples range from −9.57 to −7.20‰ and −62.49 to −46.98‰, with averages of −8.61‰ and −58.46‰, respectively. For the reservoirs in the study area, $\delta^{18}O$ and δD values range from −8.89 to −0.89‰ and

Table 5.10 Statistical results of hydrogen and oxygen stable isotopes for upstream, midstream, and downstream river water samples (‰)

	Upstream		Midstream		Downstream		Wei River		Reservoirs	
Statistics	$\delta^{18}O$	δD	$\delta^{18}O$	δD	$\delta^{18}O$	δD	$\delta^{18}O$	δD	$\delta^{18}O$	δD
Max	−7.72	−48.50	−5.75	−46.17	−6.61	−47.00	−7.20	−46.98	−0.89	−20.51
Min	−10.00	−66.84	−10.05	−68.60	−10.67	−72.94	−9.57	−62.49	−8.89	−58.17
Mean	−8.91	−57.90	−8.45	−56.38	−8.74	−58.22	−8.61	−58.46	−6.43	−45.67
Variance	0.53	3.84	1.08	6.29	0.89	5.81	0.56	3.99	2.77	12.83

−58.17 to −20.51‰, with averages of −6.43‰ and −45.67‰, respectively. The hydrogen and oxygen stable isotope relationships of upstream, midstream, downstream tributaries, the Wei River, and reservoirs in the study area are shown in Fig. 5.14. Figure 5.14 indicates that water samples from the midstream and downstream sections of tributaries as well as some reservoirs notably deviate from the regional meteoric water line, suggesting stronger evaporation effects on these waters.

In addition, the relationship between hydrogen and oxygen stable isotopes in surface water before and after rainfall in the study area is shown in Fig. 5.15. As can be seen from the figure, some of the surface water samples collected before rainfall deviate significantly from the local meteoric water line (LMWL), indicating that relatively strong evaporation occurred in the surface water before rainfall in the study area. This is consistent with the conclusion drawn from the Gibbs diagram.

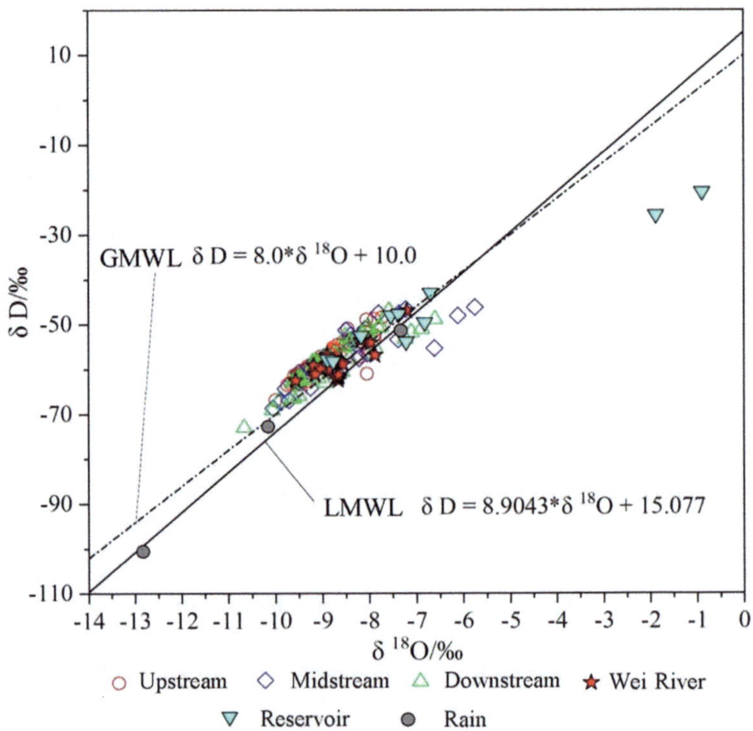

Fig. 5.14 Hydrogen and oxygen stable isotopes relationship diagram for upstream, midstream, and downstream river water samples

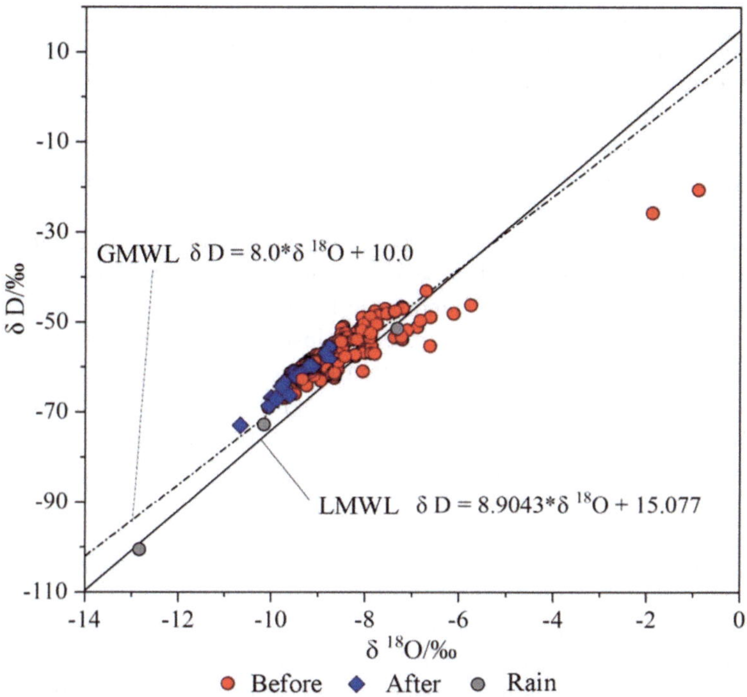

Fig. 5.15 Relationship between hydrogen and oxygen stable isotopes in surface water before and after rainfall

5.5 Ecological and Environmental Issues of Surface Water and Corresponding Measures

5.5.1 Overview of the Surface Water Ecological Environment

The Qinling Mountains hold approximately 22.2 billion m^3 of water resources within Shaanxi Province, accounting for about 50% of the province's total, and serve as the main water conservation area of Shaanxi. Among these, the northern foothills of the Qinling Mountains possess about 4 billion m^3 of water, accounting for approximately 51% of the surface water resources in the Guanzhong region. This area is a critical water conservation zone in the Wei River Basin and a key water source for the Guanzhong urban agglomeration. It is also the lifeline and core of the entire Guanzhong water ecosystem, playing a vital role in sustaining the ecological water system, ecological water volume, and the region's sustainable economic development. Historically, the northern foothills of the Qinling Mountains were characterized by lush mountains, clear waters, abundant vegetation, fertile lands, and a sound ecological environment. However, due to the long-term development of agrarian civilization, excessive human exploitation, and climate change, the ecological environment has gradually deteriorated. In recent years, driven by increasing demands for large-scale, high-yield, and high-quality agriculture and industry, shifts in consumption concepts and capacities, and the continuous development of tourism resources and commercial construction in the Qinling Mountains, as well as the overexploitation of natural and mineral resources, the ecological balance in the northern foothills has been severely disrupted. Although the Chinese government has attached great importance to environmental protection in recent years, and many departments and research institutions have made considerable efforts toward the ecological conservation of the northern Qinling region, field investigations show that while the overall ecological protection of surface water in the major valleys is relatively acceptable, many problems still persist, mainly reflected in the following aspects:

5.5.1.1 Ecological Water Shortage in Rivers, Lakes, and Wetlands

Due to the uneven spatial and temporal distribution of water resources, insufficient water allocation capacity, and encroachment by economic and social development, rivers, lakes, and wetlands in the plains are experiencing water shortages. These shortages cannot meet the ecological baseflow requirements of tributaries flowing into the Wei River, such as the Shitou River, Hei River, Lao River, Chanba River, and Ling River, as well as the ecological water needs of major lakes like Kunming Lake and Meibei Lake.

5.5.1.2 Shrinkage of Rivers, Lakes, and Wetlands

Because of reduced precipitation, ecological water deficits in rivers and lakes, overexploitation of groundwater, and other human activities, wetlands and paddy fields in the transition zones between the mountainous valleys and the plains, especially on both sides of the Ring Mountain Highway, are shrinking or have already disappeared. The surface area of rivers and lakes has decreased by about one-fifth since the 1980s. Investigations show that river flows have generally declined sharply, and many rivers experience full or partial flow interruption along their courses.

5.5.1.3 Water Environment Pollution

Field investigations reveal that many valleys still contain numerous factories. Untreated industrial wastewater, gas emissions, and solid waste are being discharged indiscriminately, posing significant pollution risks to the water environment. Downstream from the valleys, in the Xi'an and Weinan areas, surface water quality is generally classified as Class V, indicating heavy pollution. The main pollutant is ammonia nitrogen.

5.5.1.4 River Channel Pollution from Domestic Waste

The investigation found that river channels in the studied valleys suffer from severe pollution caused by domestic waste. This not only affects the visual quality of the rivers but also risks leachate contaminating surface water. In towns and villages along the rivers, untreated domestic sewage is sometimes directly discharged into rivers. In some cases, construction waste obstructs river flow. Alarmingly, in some surface water source areas, poor management has led to large amounts of household waste in river channels, waste piling up along the riverbanks, and toilets being built right next to the water, all of which pose potential threats to drinking water safety.

5.5.1.5 Vegetation Destruction, Soil Erosion, and Other Ecological Problems

Illegal and excessive mining by enterprises has led to random excavation and dumping of waste rock and slag. Road construction often results in indiscriminate disposal of excavated soil and debris. Additionally, farming on steep slopes has caused widespread vegetation destruction and soil erosion. This investigation found evidence of quarrying or mining operations in some valleys, resulting in partial damage to mountain landscapes.

5.5.2 *Major Ecological and Environmental Issues*

Due to the inherent characteristics of the aquatic ecological environment, historical deficiencies, and the anticipated increases in population, socioeconomic development, and acceleration of modernization, ecological and environmental pressures continue to intensify. The situation remains severe and complex, making water environment management a daunting challenge, water ecological restoration an arduous task, and revealing persisting shortcomings in water infrastructure and spatial coordination and control systems in need of further improvement.

5.5.2.1 Unfavorable Water Resource Endowment and Prominent Eco-environmental Issues

Influenced by climate and precipitation, water resources in the northern foothills of the Qinling Mountains are unevenly distributed in both time and space, with abrupt fluctuations in river flow, increased runoff variability, reduced total water volume, a rise in seasonal rivers, decreased water resource availability, and significant challenges in resource development and utilization. Annually, 80% of surface water is lost in the form of floods.

With the implementation of national strategies and increasing population, along with growing public aspirations for a better life and the acceleration of urbanization and industrialization, demand for water resources will continue to grow, intensifying water shortages.

Long-term severe water shortages have led to encroachment upon ecological water use, reducing river and lake surface areas by about one-fifth since the 1980s. This has caused inadequate space for aquatic ecosystems, degradation of wetland functions, and an overall low self-restoration capacity in rivers and lakes. As a result, ecological baseflows downstream of mountain outlets are hard to maintain, with frequent interruptions during dry seasons. Some rivers even experience perennial flow disruption, reducing river self-purification capacity and environmental carrying capacity. River channels are shrinking and have essentially lost their flood regulation, water supply, and ecological functions. The natural morphology of the rivers has been severely damaged, and the original landscape on both banks has deteriorated, resulting in an imbalanced aquatic ecological environment.

Some river sections have been channelized, covered, hardened, encroached upon, or filled in, causing the original riverbeds to narrow, deepen, or even disappear. Some have been converted into domestic sewage discharge channels. Consequently, some tributaries suffer from poor water quality. For instance, black and odorous water bodies are commonly observed in both urban and rural residential areas, and eutrophication in lakes and reservoirs remains uncontrolled. Additionally, overextraction of groundwater in some cities has led to the formation of groundwater depression cones, causing environmental geological issues such as land subsidence, ground fissures, and sinkholes, thereby exacerbating urban waterlogging.

5.5.2.2 Intensified Conflict Between Development and Conservation, Escalating Ecological Pressures

Over the past three decades, the ecological landscape of the northern Qinling Mountains has changed dramatically, with urban areas increasing by more than 30%, especially in Xi'an, Baoji, and Weinan, where both urban area and population have grown significantly. The dominant development model remains expansion through land occupation, resulting in spatial imbalances between regional socioeconomic growth and ecological protection. Disconnection persists between space and time, development, and natural endowment, with severe overloading of natural resources and the environment.

Rapid growth in land use for non-agricultural projects has encroached upon forests, grasslands, ponds, wetlands, and paddy fields, reducing water conservation capacity and significantly decreasing water bodies. This has locally damaged water ecology and the water environment.

Particularly problematic are instances of illegal mining, deforestation, unauthorized construction, and unregulated waste discharge in mountain valleys and surrounding zones. Tourism and agritainment developments are growing chaotically, while pollution, soil erosion, and other issues pose serious threats to ecological protection and water source security.

5.5.2.3 Persistent Shortcomings in Water Infrastructure and an Underdeveloped Water Security System

The central Shaanxi water supply network, led by the Han-to-Wei River Water Diversion Project, is not yet fully established. The limited capacity to interconnect and regulate water sources hampers supply security. On the other hand, urban expansion has encroached upon river and lake spaces, aggravating problems such as river fragmentation, reduced hydrological connectivity, and impeded flood discharge.

These issues result in low regional capacities for water resource regulation and control, exacerbating the supply-demand conflict. In particular, the region lacks flood retention zones and control infrastructure, and floodwater resources in the northern Qinling area remain underutilized. Communication and connection among lakes, between rivers and lakes, and among water systems are insufficient and unreasonable.

Furthermore, the incomplete flood control and drought mitigation system stems from the fact that comprehensive management of the southern mountain tributaries remains unfinished, with existing flood control standards still too low and posing significant safety risks. Especially with the expansion of urban scale and volume, the original flood control standards for agricultural areas along the river are far from meeting urban requirements, placing continued pressure on flood safety.

5.5.3 Main Measures to Be Taken for Ecological Environment Protection

With social development and continuous improvement in people's living standards, a good ecological environment has become an important component of quality of life and increasingly prominent in public happiness indices. Based on the region's resource advantages and characteristics, and adhering to the concept of harmonious coexistence between humans and water, while prioritizing conservation and protection, building a healthy water ecological civilization is both a livelihood issue and a public demand. Therefore, to promptly restore and protect the surface water ecological environment in the study area, a series of measures should be formulated according to the current status of the surface water ecological environment. These measures mainly include the following five aspects:

5.5.3.1 Legal Policies

In response to the ecological environment issues of surface water in Qinling, various departments in Shaanxi Province have successively issued relevant laws and regulations. These laws and regulations play a role in the water ecological environment of the northern Qinling area in two main aspects: first, dividing the northern Qinling into three zones, i.e., absolute protection zone, general protection zone, and ecological control zone, with water source conservation areas as a key basis for zoning; second, providing ecological improvement strategies or guidance for ecological development tailored to different degrees of river degradation. It is also recommended to timely and appropriately raise surface water environmental standards, implement detailed regulations on water resource utilization, management, and protection, and enforce strict rules on water use, water intake, wastewater treatment, and pollutant discharge in residential life, industrial production, and agricultural activities. This will promote green agricultural and industrial production as well as sustainable living and consumption, thereby effectively improving the surface water ecological environment. In addition, strengthening protection of water source areas is also necessary.

5.5.3.2 Regional Cooperation Mechanism

The rivers and reservoirs in the northern Qinling area are mostly managed jointly by the cities of Xi'an, Baoji, and Weinan. Water resources are shared among these cities, and flood risks are common to them as well. Therefore, it is recommended to focus on core principles such as source governance, pollution payment and compensation, and prevention of pollution transfer, emphasizing regional cooperation mechanisms to ensure the prevention, control, and sustainable utilization of surface water resources in the study area.

5.5.3.3 Economic Regulation

Economic regulation is an important tool for protecting and managing the surface water ecological environment. Key economic measures include setting tap water prices, levying ecological taxes and sewage discharge fees, and offering tax reductions to private sewage treatment enterprises. Specifically, different taxes and fees can be determined based on pollution types—for indirect pollution or waste discharge, sewage discharge fees can be imposed, and for direct pollution or waste discharge, pollution fees can be levied. These measures promote the improvement of the surface water ecological environment in the study area.

5.5.3.4 Public Participation

The public is both the main consumer and polluter of surface water resources and the primary target of policies and laws. Public awareness can be enhanced through education, publicity, announcements, and other means to communicate water resource protection policies and pollution prevention measures, thereby raising environmental protection consciousness.

5.5.3.5 Strengthening Scientific Research Related to Qinling Water Ecological Environment

To address surface water ecological problems in the northern Qinling, it is necessary to fully utilize the ecological advantages of the area, comply with local natural and socio-economic development laws, and adhere to the principle of combining protection with restoration. Solutions should be tailored scientifically and reasonably to the local conditions. Thus, government departments such as provincial science and technology, natural resources, and environmental protection should establish special scientific projects focused on Qinling's water ecological protection and restoration.

For pollution remediation and other ecological issues in the study area's surface water, it is essential to build a river inspection system and a big data platform to assess river ecological status, evaluate comprehensive carrying capacity, and monitor river management levels. A thorough investigation of pollution types and levels across rivers will advance restoration and evaluation in a comprehensive, in-depth, and multi-dimensional manner. This will realize rational utilization, restoration, and protection of surface water resources in northern Qinling, thereby optimizing environmental management and improving environmental conditions. Additionally, an ecological restoration system for the watershed should be established, with phased studies addressing different stages of water restoration.

5.5.3.6 Other Control Measures

Increase investment in sewage treatment equipment and technology development to improve treatment rates and reduce pollutant concentrations and types entering water bodies; Separate rainwater and sewage systems to reduce sewage treatment volume; adopt no-till and shallow tillage farming practices to reduce soil disturbance and nutrient loss; Reduce chemical fertilizer use in farmland, increase organic fertilizer use, and rationally apply pesticides to control agricultural non-point source pollution at the source; Improve water body structure and connectivity by installing fish passages and enhancing upstream-downstream connectivity to maintain aquatic biodiversity and food chain integrity, ensuring river ecosystem health; Continuously improve surface water ecological restoration techniques by adopting new technologies such as microbial ecological filter beds and remote sensing monitoring to compensate for limitations of traditional restoration methods.

5.6 Summary

This chapter conducted hydrological and aquatic ecological studies on 30 rivers originating from the northern Qinling Mountains through field surveys, data collection, sampling and analysis, isotope testing, and laboratory analysis. The main conclusions are as follows:

1. Based on a thorough understanding of the natural geography and human conditions of the northern Qinling Mountains, combined with survey results, the basic characteristics of surface water chemistry were comprehensively analyzed using the Piper trilinear diagram. Water quality assessments were conducted for different uses, including drinking water, irrigation water, and industrial water. The results indicate that the drinking water quality in the area is good, with Class I, II, and III water samples suitable for human consumption accounting for 54.19% of the total samples. The irrigation water quality is relatively good; according to single-factor indicators and a combined assessment considering both salinity and alkalinity hazards, the surface water in the study area is generally suitable for irrigation. The industrial water quality is also good, with water samples basically suitable for boiler use accounting for 84.5% of the total samples. River water quality shows a spatial deterioration trend from upstream to downstream and from the western to eastern parts of the study area.
2. By comprehensively utilizing methods such as the Gibbs diagram for water quality formation mechanisms, ion correlation analysis, water quality cluster analysis, and hydrogen and oxygen stable isotope analysis, the formation mechanisms and influencing factors of the surface water environment in the study area were analyzed. The formation and evolution of the surface water environment in the study area are jointly influenced by natural processes and human activities, which are intertwined and collectively control the formation and evolution of the surface water environment in the northern Qinling Mountains. Among the natural factors, water-rock interaction is the primary influence on the formation and evolution of the surface water environment, while evaporation also has a certain degree of impact. Evaporation mainly occurs in the middle and lower reaches of the tributaries flowing into the Wei River from various valleys within the study area, as well as in surface waters within the Weinan region. The human activities affecting the formation and evolution of the surface water environment in the study area are mainly agricultural activities, domestic sewage, and industrial wastewater.

3. Through field investigations of the surface water systems and surrounding environments in the study area, along with extensive literature review, the ecological environment issues of the main valley surface waters in the northern Qinling Mountains were studied. It was found that the overall ecological protection of surface waters in the main valleys of the northern Qinling is relatively good; however, there are phenomena of water shortage and wetland shrinkage within river and lake wetland ecological zones, and some rivers suffer from varying degrees of water pollution. Additionally, vegetation destruction and soil erosion are prominent issues in the study area. The main causes of these ecological problems lie in both natural and anthropogenic factors. Natural factors include climate influences leading to uneven spatial and temporal distribution of water resources, which causes water shortages and wetland shrinkage. Human factors mainly arise from large-scale tourism resource development in recent years, resulting in conflicts between development and protection, which in turn cause water pollution, soil erosion, and other ecological problems. Therefore, to further protect the surface water ecological environment of the northern Qinling and to promptly restore the degraded ecological conditions, relevant departments in the study area should strengthen efforts in five aspects: legislation and scientific research, regional cooperation mechanisms, economic regulation, public participation, and other specific management and control measures.

References

Chen J (2014) Quantitative assessment of the impacts of human activities and natural factors on river water quality and establishment of nutrients criteria in a watershed. PhD thesis, Zhejiang University. (In Chinese with English Abstract)

Gao J (2022) Research on the water ecological health evaluation of typical rivers in the Weihe River Basin (Shaanxi Section). Master thesis, Chang'an University. (In Chinese with English Abstract)

Ma Z, Li T, Hu W, Dou H (2011) Analysis of sustainable exploitation and utilization of water resources in northern Qinling. Northwest Geol 44(2):165–169. (In Chinese with English Abstract)

Nsabimana A, Li P, Alam SMK, Fida M (2023) Surface water quality for irrigation and industrial purposes: a comparison between the south and north sides of the Wei River Plain (northwest China). Environ Monit Assess 195(6):696. https://doi.org/10.1007/s10661-023-11263-0

Pang Z, Kong Y, Li J, Tian J (2017) An isotopic geoindicator in the hydrological cycle. Procedia Earth Planet Sci 17:534–537. https://doi.org/10.1016/j.proeps.2016.12.135

Salerno F, Gaetano V, Gianni T (2018) Urbanization and climate change impacts on surface water quality: enhancing the resilience by reducing impervious surfaces. Water Res 144:491–502

Stringer LC, Mirzabaev A, Benjaminsen TA, et al. (2021) Climate change impacts on water security in global drylands. One Earth 4:851–864. https://doi.org/10.1016/j.oneear.2021.05.010

Uddin MG, Nash S, Olbert AI (2021) A review of water quality index models and their use for assessing surface water quality. Ecol Indic 122:107218. https://doi.org/10.1016/j.ecolind.2020.107218

Wang D, Li P, He X, He S (2023) Exploring the response of shallow groundwater to precipitation in the northern piedmont of the Qinling Mountains, China. Urban Clim 47:101379. https://doi.org/10.1016/j.uclim.2022.101379

Wu J, Li P, Wang D et al (2019) Statistical and multivariate statistical techniques to trace the sources and affecting factors of groundwater pollution in a rapidly growing city on the Chinese Loess Plateau. Hum Ecol Risk Assess 26:1603–1621. https://doi.org/10.1080/10807039.2019.1594156

Yan T, Shen S, Zhou A (2022) Indices and models of surface water quality assessment: review and perspectives. Environ Pollut 308:119611. https://doi.org/10.1016/j.envpol.2022.119611

Zhang L, Li P, He X (2022) Interactions between surface water and groundwater in selected tributaries of the Wei River (China) revealed by hydrochemistry and stable isotopes. Hum Ecol Risk Assess 28:79–99. https://doi.org/10.1080/10807039.2021.2016054

Chapter 6
Hydrochemistry and Quality of Groundwater

Abstract Groundwater plays an irreplaceable role in ensuring the daily lives of urban and rural residents, supporting socio-economic development, and maintaining ecological balance in northern Qinling Mountains. This chapter investigates the hydrochemistry and quality of groundwater in the northern Qinling Mountains. The groundwater chemical characteristics, formation mechanisms, and quality assessment were systematically analyzed to address growing concerns over water scarcity and pollution. Results reveal that the groundwater is predominantly of the HCO_3-Ca·Mg type, with rock weathering identified as the primary controlling factor. Key findings include higher ion levels near the Wei River, and elevated nitrate in agricultural areas due to human activities. The Entropy-Weighted Water Quality Index (EWQI) evaluation indicates that 89% of samples meet the drinking water purpose, though localized pollution exists.

Keywords Groundwater quality · Hydrochemical processes · Water quality assessment · Hydrochemical types · Hydrochemical characteristics

6.1 Introduction

Groundwater chemistry serves as an ideal medium for tracing interactions between the Earth's interior and external environments, with hydrochemical characteristics providing crucial indicators of groundwater evolution patterns (Zhang et al. 2021a). Research on groundwater chemical features and their evolution mechanisms is grounded in the discipline of hydrogeochemistry. This field integrates multiple disciplines including hydrogeology, water chemistry, geochemistry, and groundwater dynamics, employing diverse methodologies to address a broad spectrum of research topics (Qian et al. 2012). Consequently, as a prominent research focus in global water science today, the spatiotemporal distribution characteristics and evolutionary mechanisms of groundwater chemistry have garnered worldwide attention.

Since the 1950s, human activities have intensified increasingly, with their impacts on groundwater chemistry becoming ever more pronounced. Hydrogeologists worldwide

P. Li et al., *Water Resources Research in Northern Qinling Mountains, China*, Springer Hydrogeology, https://doi.org/10.1007/978-3-032-09395-0_6

have conducted extensive research on the spatial evolution of groundwater chemistry using hydrochemical data. Among international scholars, Helstrup et al. (2007) investigated groundwater in the Cretaceous-Eocene limestone aquifer of southern Ghana and Togo. Employing hierarchical cluster analysis, they identified the key processes governing hydrochemical characteristics which included weathering dissolution of carbonates and silicates, mixing effects, and ion exchange. Separately, Belkhiri et al. (2012) examined the geochemical evolution of groundwater in alluvial aquifers. Their results demonstrated that in groundwater with lower electrical conductivity (EC), ionic concentrations followed the sequence: $Ca^{2+} > Na^{+} \approx Mg^{2+} > K^{+}$ and $HCO_3^- > Cl^- > SO_4^{2-} > NO_3^-$. Furthermore, as EC increases, the hydrochemical type of groundwater gradually transitions from Cl·HCO_3-Ca to Cl-Ca·Na. The primary controlling mechanisms behind these changes are the dissolution and precipitation of evaporites, coupled with ion exchange processes. Argamasilla et al. (2017) integrated multiple hydrogeochemical indicators and isotopic data to analyze the controlling factors of groundwater salinization and hydrogeochemical processes in the Marbella-Estepona coastal aquifer of southern Spain. Their findings revealed that the groundwater geochemical characteristics were predominantly governed by source rock lithology. Specifically, the hydrochemistry of eastern groundwater was primarily influenced by carbonate dissolution and rainwater leaching, whereas central-western groundwater was mainly controlled by surface water interactions. Zhang et al. (2021b) employed multivariate statistics to classify hydrochemical facies in the Jiaokou Irrigation District. Their results demonstrated that groundwater in this region could be categorized into three distinct types: HCO_3-Na, SO_4·Cl-Na and Cl·SO_4-Na, with evaporation concentration being the dominant controlling mechanism. Zhang et al. (2021c) analyzed the hydrochemical characteristics and formation-evolution mechanisms of karst groundwater in the Sangu Spring Basin. The results showed that the low total dissolved solids (TDS) water was mainly of the HCO_3-Ca·Mg type, while the high TDS water was mainly of the HCO_3·SO_4-Ca·Mg type and SO_4·HCO_3-Ca type. This hydrochemical differentiation was primarily controlled by factors including water-rock interactions, anthropogenic activities, and leakage from local fissure water systems. In summary, the analysis of groundwater chemical characteristics and their evolutionary patterns provides valuable theoretical support for regional water environment protection efforts.

At present, the methods for studying the evolution of groundwater environment can be roughly summarized as follows:

1. **Geostatistical method**: Geostatistics is an interdisciplinary subject of statistics and geology. Geostatistics was initially developed to solve problems in the entire process related to mining, with its core being the "Kriging" interpolation. The geostatistical method was initially applied in fields such as geology, soil, and meteorology, and later was also used by many scholars in groundwater research. For example, Hoogland et al. (2010) used the geostatistical method to evaluate the changes in groundwater level depth across the Netherlands and put forward suggestions for improving the accuracy of future predictions. Bradai et al. (2016) took groundwater salinity as the research object and improved the Kriging method, finding that compared with the ordinary Kriging method, the improved

indicator Kriging method had better spatial predictability in determining the salinity area. Osiakwan et al. (2021) used a comprehensive method combining mathematical statistics, geostatistical analysis, and information entropy to characterize and identify the main factors and processes controlling groundwater quality in coastal aquifers. The results showed that the salinity varied greatly from the coast to the inland.

2. **Hydrogeochemical method**: The groundwater environment is affected not only by the recharge sources, but also by various factors such as the mineral composition of aquifer media and geological structure during the runoff process. During the groundwater recharge-runoff-discharge process, hydrochemical characteristics can well reflect the occurrence environment, runoff path and hydrochemical processes of groundwater. As early as the beginning of the twentieth century, hydrochemical analysis had been applied to research related to the water environment. Later, a number of graphical methods such as Gibbs diagram, Schoeller diagram, Piper diagram and USSL diagram emerged to reflect the variation characteristics and mechanisms of hydrochemical components (Gibbs 1970; Piper 1944). At present, scholars generally combine multivariate statistics, graphical methods, ionic ratio, hydrogeochemical simulation and other means to analyze the dynamic changes of groundwater flow and the temporal and spatial variations of hydrochemistry. For example, Zhang et al. (2021a, b) used correlation analysis, principal component analysis, saturation index, Gibbs diagram, Piper diagram and ion ratio to analyze the current situation of fluoride pollution, hydrochemical components, types and their formation causes of groundwater in the Jiaokou Irrigation District. The results revealed that the groundwater chemical field in the study area was not only related to recharge conditions such as rainfall and long-term irrigation, but also largely affected by evaporation and concentration effects. Jia et al. (2020a) analyzed the hydrochemical evolution of groundwater in the central area in the Guanzhong Basin by combining multivariate statistical methods, ionic ratio methods and Durov diagram. The results showed that agricultural activities had a great impact on the hydrochemical characteristics of groundwater in the study area. Compared with precipitation, the groundwater was more derived from irrigation water and leakage from canals and reservoirs. Chen et al. (2021) comprehensively analyzed the spatial distribution characteristics of fluoride in groundwater in the Guanzhong Basin by multivariate statistics, Gibbs diagram and ion exchange, finding that high-fluoride groundwater was mainly distributed in the area north of the Wei River and east of the Qishui River, while high pH, low Ca^{2+}, high HCO_3^- and Na^+ were the main factors causing high fluoride groundwater.
3. **Isotopic method**. The application of isotopes in hydrogeology benefits from the development of mass spectrometers. The fractionation and decay characteristics of isotopes were used to study the origin, formation, and changes in the quality and quantity of groundwater and to identify the recharge intensity, age, flow direction, and velocity of groundwater (Qian et al. 2012). Compared with other methods, the isotope hydrology method has more advantages in studying the groundwater environment, and thus has gradually become one of the important

research methods for the research of groundwater environment evolution. For example, Schuster et al. (2003) used stable isotopes δ^2H and $\delta^{18}O$ to explore the mutual evolution relationship between Williams Lake and groundwater in Minnesota, and found that half of the lake water come from local groundwater, and groundwater also input most chemical substances while inputting water into the lake. Farid et al. (2013) analyzed the chemical and isotopic characteristics of groundwater, rainwater, and surface water in the Paleogene-Neogene strata of Ain Bou Mourra, and found that surface water and groundwater had strong spatial variability, modern rainfall significantly recharged groundwater, while surface water was significantly affected by evaporation. Jia et al. (2020b) used hydrogen, oxygen, nitrogen, and other isotopes to identify the sources of nitrogen pollution in shallow groundwater in the northern Qinling Mountains. The results showed that the Jijiahe Reservoir and Shidi River in the center were the main sources of aquifer recharge; in addition, the results of $\delta^{15}N\text{-}NO_3^-$ and $\delta^{18}N\text{-}NO_3^-$ indicated that although the nitrate concentration in rivers and reservoirs was relatively low at present, the high-concentration nitrogen pollutants released by local fertilizer factories could be potential pollution sources in the future.

This chapter uses hydrochemical and isotopic characteristics to understand the sources, formation mechanisms, and water quality of groundwater in the northern Qinling Mountains. The main objectives of this chapter are (1) to reveal the hydrochemical characteristics of groundwater in the northern Qinling Mountains, (2) to identify the sources and processes controlling of groundwater geochemistry and (3) to evaluate the groundwater quality in the study area. The results are instrumental for regional groundwater quality protection and management.

6.2 Hydrochemical Characteristics of Groundwater

6.2.1 Sampling

Field hydrological and ecological environment surveys were conducted throughout the entire fieldwork phase of the project, covering a vast piedmont plain area from the northern Qinling Mountains to the Wei River, with a total area of 14,600 km^2. In this field survey, 119 groundwater samples were collected. The distribution map of groundwater sampling points is shown in Fig. 6.1. Three samples were taken from each water sampling point: two of them were placed in a 500 mL sampling bottle and a 2.5 L polyethylene plastic bucket for determining the content of conventional physical and chemical indicators. The sampling, preservation, and transportation of all samples complied with national guidelines. The samples collected from the field survey were sent to the Laboratory of Xi'an Mineral Resources Survey Center for testing indicators such as K^+, Na^+, SO_4^{2-}, and Cl^-; to Shaanxi Nuclear Industry Group Comprehensive Analysis and Testing Co., Ltd. for testing indicators

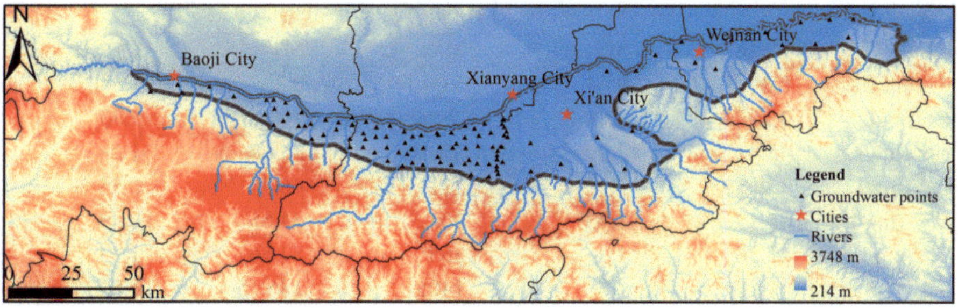

Fig. 6.1 Distribution of groundwater sampling points

including F^-, I, Mn, Fe, and Cr^{6+}; to Shaanxi Keyi Yangguang Testing Technology Service Co., Ltd. for testing 38 conventional drinking water indicators such as total coliforms, Hg, and As. On-site testing indicators included water temperature, pH, electrical conductivity (EC), dissolved oxygen (DO), and redox potential (ORP).

6.2.2 Hydrochemical Characteristics

Statistical analysis was conducted on the groundwater samples in the study area. The analyzed indicators include, in addition to pH, TDS, total hardness (TH), and the eight conventional ions (K^+, Na^+, Ca^{2+}, Mg^{2+}, Cl^-, SO_4^{2-}, HCO_3^-, CO_3^{2-}), F^-, three nitrogen forms (NO_3^-, Pb, NH_4^+), chemical oxygen demand, and heavy metal indicators (Fe, Cu, Mn, Zn, Hg, Cr^{6+}, As, Pb, Cd). The statistical results of the major ions in groundwater are shown in Table 6.1.

1. pH, TDS, and TH
   As shown in Table 6.1, the pH values of groundwater in the northern Qinling Mountains in 2020 ranged from 6.83 to 9.12, with an average of 7.62, indicating that the groundwater in the study area is weakly alkaline. The coefficient of variation (CV) for pH was very small (0.04), indicating weak variability. TDS reflects the amount of dissolved substances in water. Groundwater can be classified into three categories based on TDS values: Freshwater: TDS < 1000 mg/L; Slightly saline water: 1000–3000 mg/L; Saline water: TDS > 3000 mg/L. In 2020, TDS values in groundwater fluctuated between 120 and 21,340 mg/L, with an average of 700 mg/L, suggesting that most groundwater samples were freshwater. The CV for TDS was 0.72, indicating moderate variability. TH refers to the total concentration of calcium and magnesium ions in groundwater, reflecting its hardness. Prolonged consumption of soft water (very low TH) may increase the risk of cardiovascular diseases, while hard water (high TH) can cause gastrointestinal disorders and potentially lead to kidney stones. According to the hardness classification standard (expressed as $CaCO_3$), groundwater can be divided into four grades: (soft water: TH < 150 mg/L), (slightly hard water:

Table 6.1 Statistical characteristics analysis of groundwater chemistry in 2020

Items	Mean	Min	Max	SD	CV
F^- (mg/L)	0.29	0.04	5.00	0.46	1.58
K^+ (mg/L)	3.22	0.48	76.60	7.48	2.33
Na^+ (mg/L)	45.41	2.29	308.00	55.68	1.23
Ca^+ (mg/L)	120.58	20	581	66.78	0.55
Mg^+ (mg/L)	35.12	2.43	204	34.65	0.99
Cl^- (mg/L)	43.26	1	365	57.56	1.33
SO_4^{2-} (mg/L)	116.04	4.8	1405	186.15	1.60
HCO_3^- (mg/L)	332.93	48.8	830.0	134.84	0.41
NO_3^- (mg/L)	21.23	0.00	171.39	28.42	1.34
TH (mg/L)	435.01	65.1	2027	274.66	0.63
TDS (mg/L)	699.95	120	3828	506.47	0.72
pH	7.62	6.83	9.12	0.34	0.04

150–300 mg/L), (hard water: 300–450 mg/L) and (extremely hard water: TH > 450 mg/L). In 2020, the TH of groundwater in the northern Qinling Mountains showed great concentration variations, with the minimum value being only 65.1 mg/L, the maximum value reaching as high as 2027 mg/L, and the average value being 435.01 mg/L, indicating that the some of the groundwater is not suitable for human consumption due to its high hardness. The coefficient of variation of TH is 0.63, indicating moderate variability.

2. Major Cations
   The maximum, minimum, and average K^+ concentrations in groundwater were 76.6 mg/L, 0.48 mg/L, and 3.22 mg/L, respectively, indicating low K^+ levels. Na^+ concentrations ranged from 2.29 to 308 mg/L, with an average of 45.41 mg/L. The CV for Na^+ was 1.23, indicating strong variability. Ca^{2+} and Mg^{2+} are key indicators of water hardness. In 2020, Ca^{2+} concentrations ranged from 20 to 581 mg/L (mean 120.58 mg/L), while Mg^{2+} concentrations ranged from 2.43 to 204.00 mg/L (mean 35.12 mg/L). The CVs for Ca^{2+} and Mg^{2+} were 0.55 (moderate variability) and 0.99 (strong variability), respectively.
3. Major Anions
   Among the major anions, Cl^- concentrations ranged from 1 to 365 mg/L (mean 43.26 mg/L), with a CV of 1.33, indicating strong variability. SO_4^{2-} concentrations ranged from 4.8 to 1405 mg/L (mean 116.04 mg/L). Long-term consumption of groundwater with high SO_4^{2-} concentrations may cause gastrointestinal disturbances such as diarrhea and dehydration. The CV for SO_4^{2-} also indicated strong variability, suggesting significant spatial heterogeneity in both Cl^- and SO_4^{2-} distributions. HCO_3^- concentrations were relatively high and varied widely, ranging from 48.8 to 830 mg/L (mean 332.93 mg/L). The CV for HCO_3^- was 0.41 (moderate variability).

4. F^- and NO_3^-
   Fluoride is essential for human health, but both deficiency and excess can be harmful. Optimal fluoride levels in groundwater are defined as 0.5–1.5 mg/L. In 2020, fluoride concentrations in the study area ranged from 0.40 to 5.00 mg/L, with an average of 0.29 mg/L. The CV for F^- was 1.58%, indicating strong variability. Nitrogen pollution (nitrate, nitrite, and ammonia) is a common issue in groundwater affected by intensive agricultural activities. In 2020, NO_3^--N concentrations in the study area ranged from 0 to 171.3 mg/L (mean 21.23 mg/L). The maximum value exceeded the national standard drinking water limit (20 mg/L as N) by 8.5 times. Prolonged consumption of high-NO_3^- groundwater may cause methemoglobinemia (blue baby syndrome), gastric diseases, thyroid dysfunction, reproductive issues, and increased risk of miscarriage. The CV for NO_3^- was 1.34, indicating strong spatial variability.
   The spatial distribution of major hydrochemical indicators in the groundwater of the study area in 2020 is shown in Fig. 6.2. Except NO_3^-, all chemical indicators

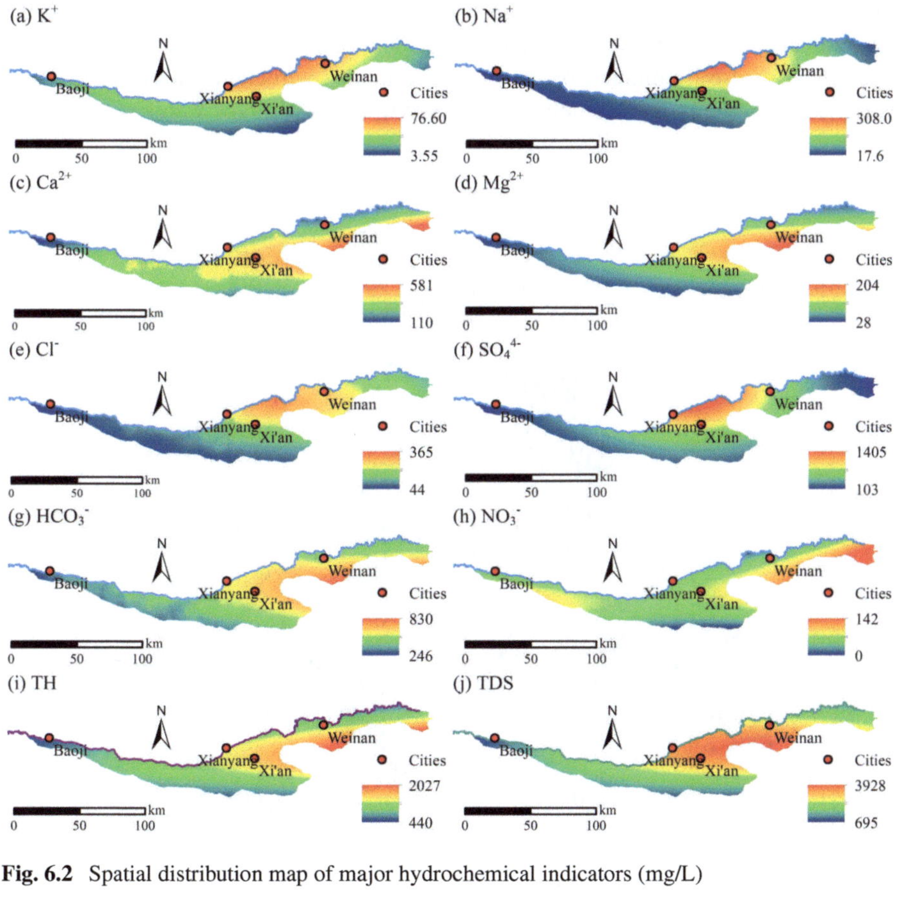

Fig. 6.2 Spatial distribution map of major hydrochemical indicators (mg/L)

exhibit a trend of being low in the east and west but high in the middle. Moreover, the closer to the Weihe River, the higher the ion content will be. NO_3^- shows a pattern of being high in the east and west but low in the middle. This is because the eastern and western parts of the northern Qinling Mountains are agricultural areas where a large number of crops such as kiwifruit, wheat, and corn are planted, while the central region is the urban agglomeration of Xi'an and Xianyang.

6.2.3 Hydrochemical Types

During groundwater flow, the combined influence of natural and anthropogenic factors generates diverse hydrochemical facies. These facies serve as diagnostic indicators of groundwater evolution patterns and governing mechanisms. Analysis of spatiotemporal variations in hydrochemical facies enables identification of groundwater flow paths and potential water-rock interactions. This study employs the Piper diagram to characterize groundwater chemical types (Fig. 6.3). Cations in the groundwater of the study area are dominated by Ca^{2+}, accounting for more than 50%. Anions are dominated by HCO_3^-. The main hydrochemical type is HCO_3-Ca·Mg type water, while a small amount is SO_4·Cl-Ca·Mg type water.

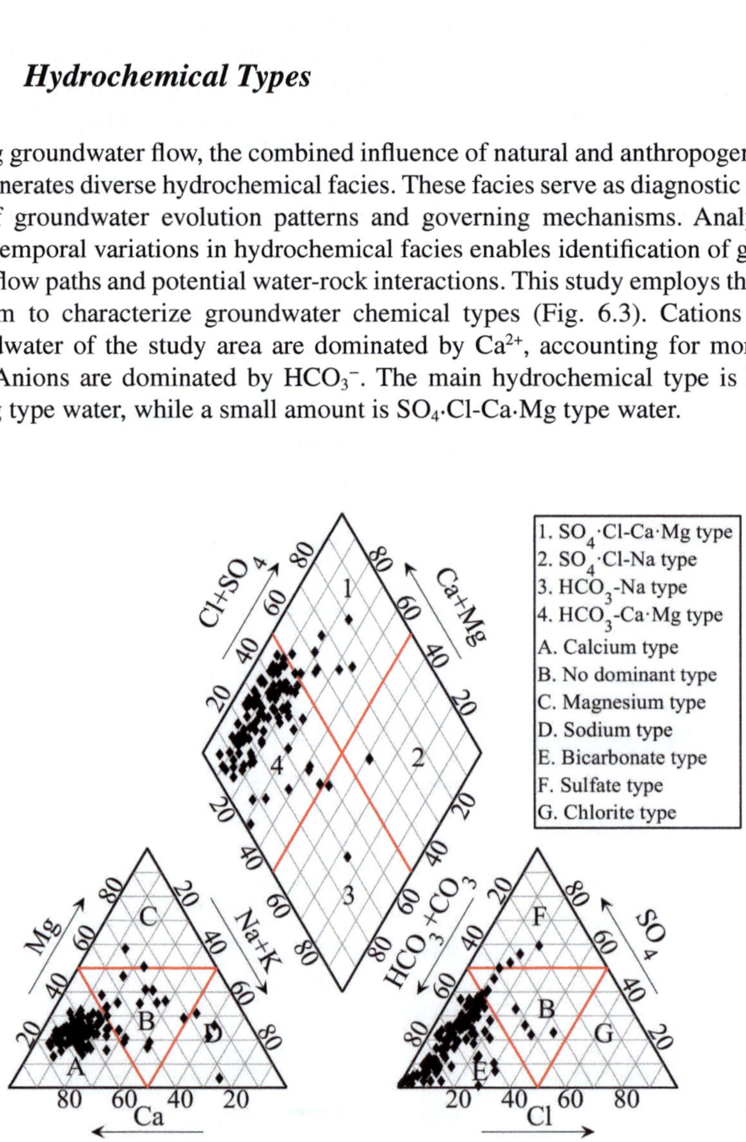

Fig. 6.3 Piper diagram of groundwater

6.3 Mechanisms of Hydrochemical Processes

Groundwater chemical composition evolves through long-term interactions among water, subsurface geology, and external environments. These hydrochemical characteristics serve as indicators of groundwater circulation processes, revealing water-rock interactions between groundwater and surrounding formations, thereby providing scientific underpinnings for understanding the formation of chemical constituents and quality evolution in groundwater systems. Building upon the comprehensive characterization of hydrochemical facies in the northern Qinling Mountains, this section further investigates the sources and formation mechanisms of groundwater chemical constituents.

6.3.1 *Gibbs Diagram*

In 1970, based on the analysis of hydrochemical components in a large number of water bodies such as rainwater, river water, lake water and seawater, Gibbs proposed that the hydrochemical components of natural water bodies are mainly controlled by three main end-members: atmospheric precipitation, rock weathering, and evaporation/concentration. Scatter plots (Gibbs diagrams) of $Na^+/(Na^++Ca^{2+})$ vs TDS and $Cl^-/(Cl^- + HCO_3^-)$ vs TDS can be used for qualitative analysis of the influencing factors of hydrochemical components in water bodies. Though this diagram was originally proposed for surface water, it has been widely used in the study of groundwater formation mechanisms. In this study, all groundwater samples were collected from shallow aquifers, and they have close hydraulic connection with surface water. Therefore, in this section, the Gibbs diagram was used for a preliminary qualitative analysis of the main controlling factors of groundwater chemical components in the northern Qinling Mountains. As shown in Fig. 6.4, the main controlling factor of groundwater in the study area is mainly rock weathering, and individual samples are controlled by evaporation and concentration. This indicates that water-rock interactions are the most important processes regulating the formation of groundwater hydrochemistry.

6.3.2 *Ionic Ratios*

Before analyzing the major ionic ratios of groundwater, cluster analysis was performed on 119 groundwater samples collected in 2020. The results showed that the study area could be divided into three categories using cluster analysis at a linkage distance of 50 (Fig. 6.5). The classification of these three groups of water samples is closely linked to the TDS content. The first group (C2) consists of samples with low TDS content, where HCO_3^- is the dominant anion; the third group (C3) includes

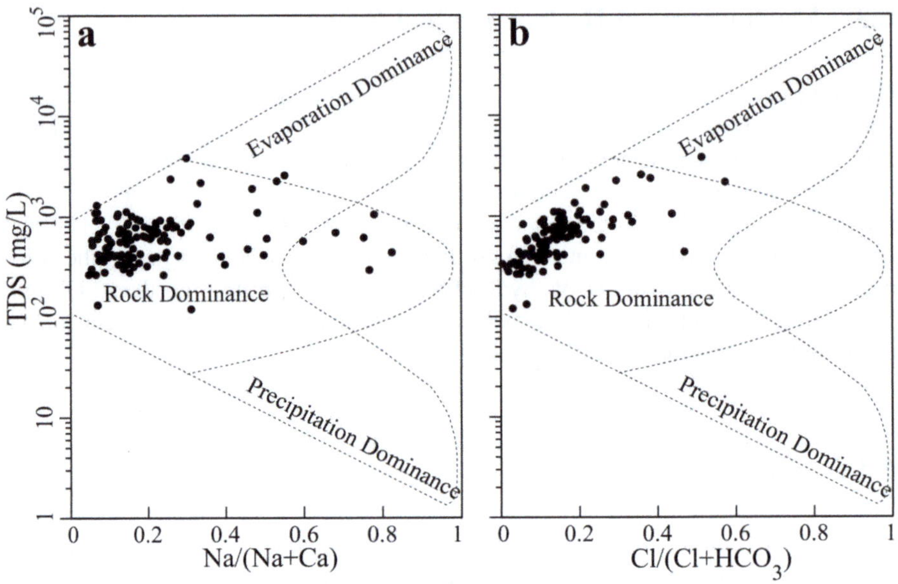

Fig. 6.4 Gibbs diagram of groundwater in the study area

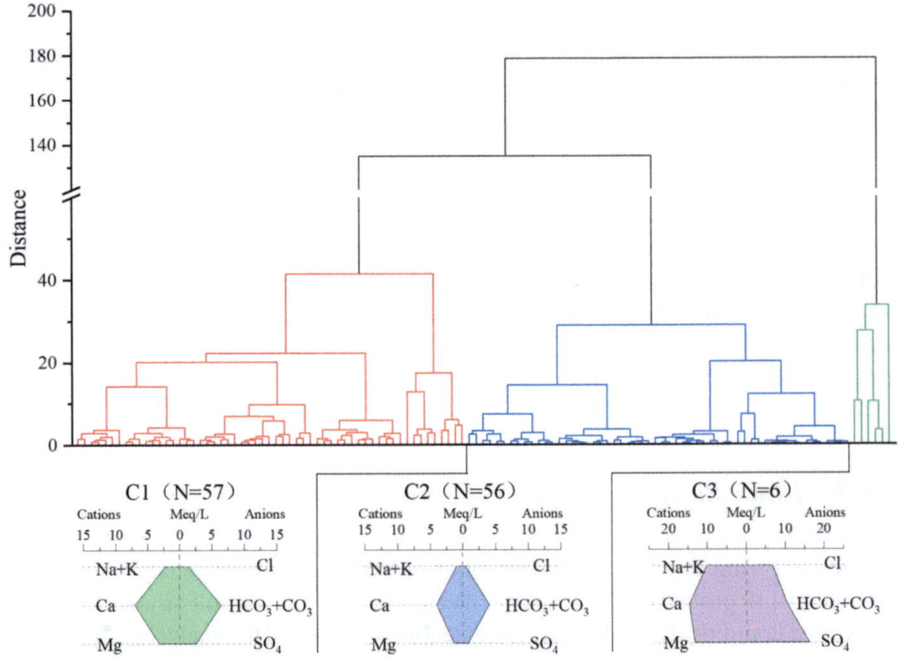

Fig. 6.5 Cluster analysis and stiff diagrams of groundwater

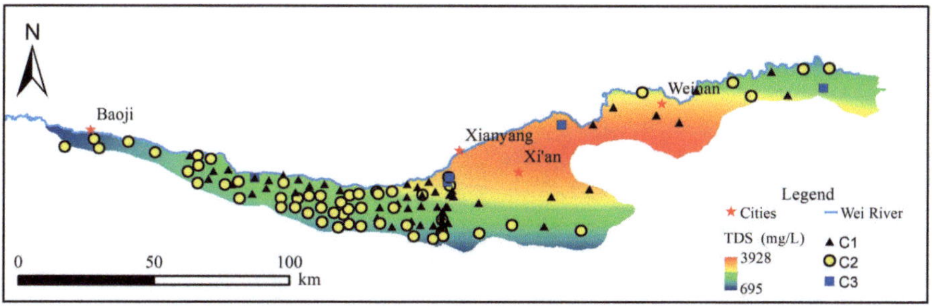

Fig. 6.6 Spatial Distribution Map of Groundwater Clustering Results and TDS in the Northern Foothills of the Qinling Mountains

samples with high TDS content, with SO_4^{2-} as the primary anion; and the group falling between these two categories is the second group (C1). Specifically, the TDS content of groundwater samples increases gradually from C2 to C1 and then to C3. As shown in the Fig. 6.6, there are 57 and 56 groundwater samples belonging to C1 and C2, respectively, both of which are of the Na-HCO_3 type. These two types of water samples are distributed throughout the study area, with the distribution of C1 being closer to the Weihe River compared to C2. There are 6 groundwater samples in C3, characterized by the Ca-SO_4 type, which are mainly distributed in the central part of the study area (the Xianyang-Xi'an-Weinan urban agglomeration region).

Based on the results of cluster analysis, the hydrogeochemical issues of the groundwater environment in the northern Qinling Mountains were analyzed using ionic ratios. The distribution characteristics of the milliequivalent concentrations of various ions (expressed in meq/L) in the study area are shown in Fig. 6.6. As shown in Fig. 6.7a, Na^+ and $C1^-$ exhibit a roughly linear distribution. Moreover, along the sequence from C2 to C1 to C3, the contents of Na^+ and Cl^- increase progressively. This is because the distribution of C2 → C1 → C3 extends from the piedmont to the vicinity of the Weihe River, where the groundwater migration time is significantly prolonged, leading to the continuous enrichment of Na^+ and Cl^- and a gradual increase in TDS content. The dissolution of halite can be expressed in Eq. (6.1). In addition, nearly all water sample points are located above the 1:1 line, indicating that Na^+ concentrations exceed those of Cl^-, suggesting that, in addition to halite dissolution, other sources such as the dissolution of albite and mirabilite (Eqs. 6.2 and 6.3) may contribute to the Na^+ content in the groundwater of the northern foothills of Qinling Mountains. In addition, cation exchange may also affect the content of Na^+ in groundwater.

$$NaCl \rightarrow Na^+ + Cl^- \tag{6.1}$$

$$\begin{aligned} &2NaAlSi_3O_8 + 2CO_2 + 11H_2O \rightarrow Al_2Si_2O_5\left(OH\right)_2 \\ &+4H_4SiO_4 + 2Na^+ + 2HCO_3^- \end{aligned} \tag{6.2}$$

$$Na_2SO_4 \rightarrow 2Na^+ + SO_4^{2-} \tag{6.3}$$

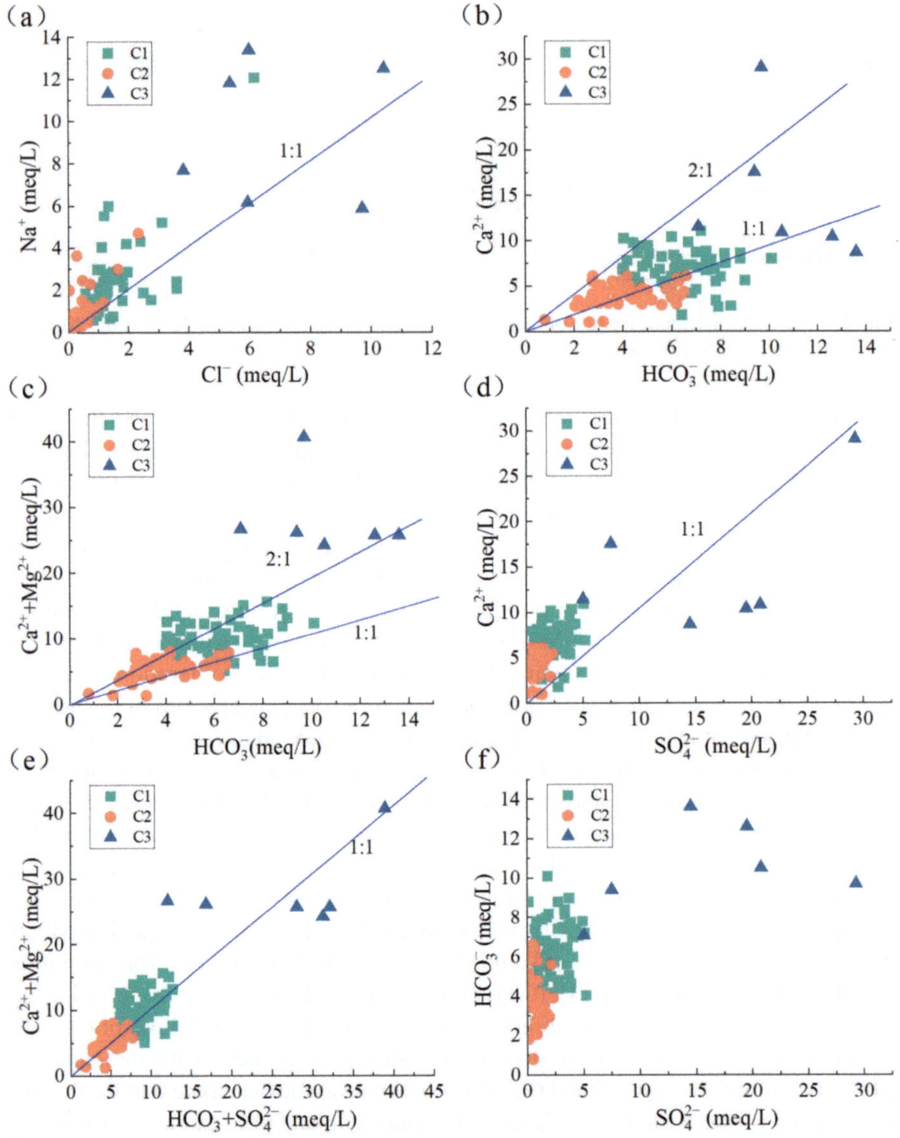

Fig. 6.7 Ionic ratios of groundwater in the northern foothills of Qinling Mountains

The dissolution of carbonates is generally the main source of Ca^{2+}, Mg^{2+}, and HCO_3^-. When CO_2 is involved in the system, the dissolution reaction equations of calcite and dolomite are specifically shown as Eqs. (6.4), (6.5), (6.6) and (6.7). The concentration ratios of HCO_3^- vs. Ca^{2+} and HCO_3^- vs. (Ca^{2+}+Mg^{2+}) resulting from the reactions should be between 1:1 and 2:1. As shown in Fig. 6.7b, most water sample points lie between the 1:1 and 2:1 lines, while a small number are located below the 2:1 line. This indicates that the Ca^{2+} content is lower than the amount generated by theoretical calcite dissolution, suggesting that although calcite dissolution is one of the sources of Ca^{2+} and HCO_3^- in the study area, cation exchange have reduced the Ca^{2+} content in the water. Similarly, Fig. 6.7c shows that most water sample points fall between the 1:1 and 2:1 lines, indicating that dolomite is also one of the sources of Ca^{2+}, HCO_3^-, and Mg^{2+} in the study area.

$$CaCO_3 + H^+ \rightarrow HCO_3^- + Ca^{2+} \tag{6.4}$$

$$CaCO_3 + H_2O + CO_2 \rightarrow 2HCO_3^- + Ca^{2+} \tag{6.5}$$

$$CaMg(CO_3)_2 + 2H^+ \rightarrow 2HCO_3^- + Ca^{2+} + Mg^{2+} \tag{6.6}$$

$$CaMg(CO_3)_2 + 2H_2O + CO_2 \rightarrow 4HCO_3^- + Ca^{2+} + Mg^{2+} \tag{6.7}$$

Gypsum, as one of the main soluble minerals in the study area, is a major source of Ca^{2+} and SO_4^{2-}. When the ratio of SO_4^{2-} to Ca^{2+} is 1:1, it indicates that Ca^{2+} and SO_4^{2-} in the groundwater of the study area mainly come from the dissolution of gypsum, with the specific reaction formula as follows (Eq.6.8):

$$CaSO_4{\cdot}2H_2O \rightarrow Ca^{2+} + SO_4^{2-} + 2H_2O \tag{6.8}$$

Figure 6.7d presents the ionic ratio diagram of SO_4^{2-} vs Ca^{2+}. It can be seen from the diagram that all water sample points are located above the 1:1 line, indicating that besides gypsum, the source of Ca^{2+} also comes from the dissolution of carbonates such as calcite. In addition, the dissolution of mirabilite and some human activities (such as the application of fertilizers in agriculture) can also increase the content of SO_4^{2-}.

Figure 6.7e shows the comprehensive effect of the dissolution of various carbonate minerals (calcite, dolomite) and gypsum in groundwater. When all three minerals are dissolved, the ratio of Ca^{2+}+Mg^{2+} vs HCO_3^- + SO_4^{2-} in groundwater should be close to 1:1. As can be seen in the figure, all water sample points are close to the 1:1 line, indicating that the main sources of Ca^{2+}, Mg^{2+}, HCO_3^-, and SO_4^{2-} in the groundwater of the northern foothills of the Qinling Mountains are the dissolution of carbonate minerals and gypsum, and these ion contents of C3 and C1 are significantly higher than those of C2. This suggests that the dissolution of carbonate minerals and gypsum in the groundwater samples of C1 and C3 is higher.

Figure 6.7f shows that the content of HCO_3^- in the study area is higher than that of SO_4^{2-}, and the three clustering results exhibit a good correlation with both HCO_3^- and SO_4^{2-}. Specifically, from C2 to C1 to C3, both HCO_3^- and SO_4^{2-} increase gradually. This indicates that the dissolution degree of calcite and dolomite in the groundwater of the northern foothills of the Qinling Mountains is higher than that of gypsum.

6.3.3 Correlation Analysis

The Pearson correlation coefficient mainly reflects the relationship between two variables, and is widely used in the field of hydrogeochemistry to analyze the ion sources. Specifically, when the correlation coefficient r^2 is greater than 0.7, there is a strong correlation between the two indicators; when r^2 value is between 0.3 and 0.7, there is a moderate correlation, and when r^2 is less than 0.3, there is a weak correlation. In this section, the heat map of the Pearson correlation coefficient matrix of each indicator was draw (Fig. 6.8).

As illustrated in Fig. 6.8, pH shows a weak correlation with most major ion indicators. Strong positive correlations exist between each pair of K^+, Na^+, Ca^{2+}, Mg^{2+}, Cl^-, SO_4^{2-}, and TDS. Specifically, TDS exhibits a strong correlation ($r^2 > 0.7$) with total hardness (TH), Na^+, Ca^{2+}, Mg^{2+}, Cl^-, and SO_4^{2-}, indicating that these ions make significant contributions to the TDS of phreatic water in the study area. Notably, Na^+ is strongly correlated with Mg^{2+}, Cl^-, and SO_4^{2-}, a phenomenon attributed to their shared material sources, these ions are typically enriched in evaporite minerals and associated with the dissolution and weathering of halite and Glauber's salt (Fig. 6.7a). Mg^{2+} demonstrates strong correlations with Cl^-, and SO_4^{2-}, while HCO_3^- also correlates strongly with Mg^{2+}. This suggests that in addition to ion enrichment caused by evaporation in phreatic water, dissolution of dolomite and magnesite is also occurring (Eqs. 6.7 and 6.9).

$$MgCO_3 + H_2O + CO_2 \rightarrow 2HCO_3^- + + Mg^{2+} \tag{6.9}$$

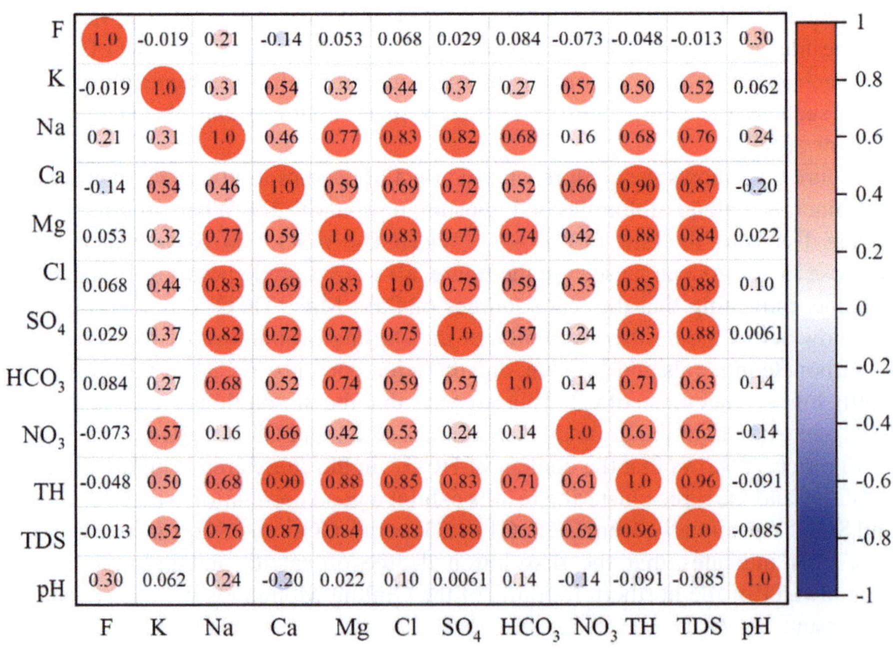

Fig. 6.8 Heatmap of Pearson correlation for various indicators of groundwater

Furthermore, NO_3^- displays moderate correlations with Cl^-, Ca^{2+}, Mg^{2+}, and K^+. This can be explained by the fact that most areas in the northern Qinling Mountains are agricultural regions. Nitrogen and potassium fertilizers are applied in these areas, introducing NO_3^- and K^+ into groundwater after being leached by rainwater and irrigation water. Moreover, during infiltration, Ca^{2+} and K^+ are more prone to synchronous migration with NO_3^- due to their weaker adsorptivity compared to Na^+ and Mg^{2+}.

6.4 Groundwater Quality Assessment

Groundwater is widely used in the northern Qinling Mountains, such as for domestic drinking, livestock drinking, agricultural irrigation, and industrial production. However, due to rapid economic development, population growth, and the development of industry and agriculture, groundwater quality has been severely polluted, and problems such as groundwater salinization, high fluoride, and high nitrogen have become increasingly prominent. Therefore, in order to realize the sustainable development of water resources, it is very necessary to carry out groundwater quality assessment. The temporal and spatial evolution laws of groundwater quality can provide a strong guarantee for the better utilization and development of groundwater resources in the region.

6.4.1 Water Quality Evaluation Methods

Groundwater quality evaluation is a quantitative description of the applicability and quality of groundwater, and it serves as an important basis for managing groundwater resources. At present, there are many methods commonly used for groundwater quality evaluation, and in this book, the Entropy-Weighted Water Quality Index (EWQI) was used for the overall groundwater quality assessment. EWQI is a comprehensive method for evaluating groundwater quality. It can consider multiple water quality indicators in the evaluation process and reflect the impact of each indicator on comprehensive water quality based on their content. The main calculation process is as follows:

1. Establish the initial matrix
   Suppose there are a total of m ($i = 1, 2, 3, \ldots, m$) water samples to be assessed, and each water sample has n ($j = 1, 2, 3, \ldots, n$) water quality indicators. The m water samples and n water quality evaluation indicators form an $m \times n$ initial matrix X. X is therefore expressed as Eq. (6.10).

$$X = \begin{bmatrix} x_{11} & x_{12} & \cdots & x_{1n} \\ x_{21} & x_{22} & \cdots & x_{2n} \\ \vdots & \vdots & \ddots & \vdots \\ x_{m1} & x_{m2} & \cdots & x_{mn} \end{bmatrix} \tag{6.10}$$

2. Standardize the initial matrix
   Due to the different orders of magnitude of various water quality parameters, they are standardized to reduce the impact caused by different orders of magnitude. In the formula (Eq. 6.11), y_{ij} represents the standardized value, $(x_{ij})^j_{\min}$ denotes the minimum value among the initial values of indicator j, and $(x_{ij})^j_{\max}$ represents the maximum value among the initial values of indicator j.

$$y_{ij} = \begin{cases} \dfrac{x_{ij} - (x_{ij})^j_{\min}}{(x_{ij})^j_{\max} - (x_{ij})^j_{\min}} \text{(Efficiency type)} \\ \dfrac{(x_{ij})^j_{\max} - x_{ij}}{(x_{ij})^j_{\max} - (x_{ij})^j_{\min}} \text{(Cost type)} \end{cases} \tag{6.11}$$

After standardizing the matrix, the standard evaluation matrix for calculation can be obtained (Eq. 6.12):

$$Y = \begin{bmatrix} y_{11} & y_{12} & \cdots & y_{1n} \\ y_{21} & y_{22} & \cdots & y_{2n} \\ \vdots & \vdots & \ddots & \vdots \\ y_{m1} & y_{m2} & \cdots & y_{mn} \end{bmatrix} \tag{6.12}$$

3. Calculate the weights of different indicators
   Let P_{ij} represent the j-th indicator of the i-th sample. To avoid y_{ij} being zero, 10^{-4} is used as a correction coefficient (Eq. 6.13).

$$P_{ij} = \frac{y_{ij} + 10^{-4}}{\sum_{i=1}^{m} \left(y_{ij} + 10^{-4}\right)} \tag{6.13}$$

Based on the results obtained from the above formula, substitute them into the equation to get the information entropy of each indicator. The calculation method is as follows (Eq. 6.14):

$$e_j = -\frac{1}{\ln m} \sum_{i=1}^{m} P_{ij} \ln P_{ij} \tag{6.14}$$

Then, the weight of each indicator is calculated through information entropy, and the calculation method is as follows (Eq. 6.15):

$$W_j = \frac{1 - e_j}{\sum_{j=1}^{n} \left(1 - e_j\right)} \tag{6.15}$$

4. Calculate the quality grade of each indicator
   The q_j is obtained by calculating the quality grade of each indicator, where c is the measured value of each indicator and S is the allowable limit of each indicators (Eq. 6.16).

$$q_j = \begin{cases} \dfrac{c_j}{S_j} \times 100 \\ \dfrac{c_{\text{pH}} - 7}{S_{\text{pH}} - 7} c_{pH} - 7 \end{cases} \tag{6.16}$$

5. Calculate the EWQI value
   The EWQI can be obtained by multiplying the weight of each indicator with its quality grade (Eq. 6.17)

$$EWQI = \sum_{j=1}^{n} W_j \times q_j \tag{6.17}$$

From the perspective of reflecting the characteristics of the groundwater environment in the study area, and considering the actual situation in the region comprehensively, a total of 7 indicators, including pH, TDS, TH, Cl^-, SO_4^{2-}, F^-, and NO_3^-, were selected as water quality evaluation indicators to assess the groundwater quality. The Standard for Groundwater Quality (GB/T 14848-2017) was adopted as the evaluation criterion to assess the groundwater quality in the northern foothills of the Qinling Mountains.

In the Chinese groundwater quality standard (standard no. GB/T 14848-2017), groundwater quality is classified into five categories, with the standard values for each indicator corresponding to each quality category listed in Table 6.2. Correspondingly, the EWQI can also be divided into five categories, as detailed in Table 6.3.

6.4.2 Water Quality Evaluation Results

The evaluation results are shown in Table 6.4. In the study area of the northern Qinling Mountains, Class I water accounts for approximately 4.2% (5 water samples), Class II water approximately 52.1% (62 water samples), Class III water approximately 32.8% (39 water samples), Class IV water approximately 8.4% (10 water samples), and Class V water accounts for the smallest proportion, approximately 2.5% (3 water samples).

The spatial distribution of the EWQI evaluation results is shown in Fig. 6.9. Class I water is sporadically distributed in the Baoji-Xianyang section. Class II and Class III water are mainly distributed in the western region (Baoji City) and central region (western Xi'an) of the northern Qinling Mountains. Class III water is

Table 6.2 Classification standards for various indicators of groundwater quality (mg/L, except pH))

Items	I	II	III	IV	V
pH	6.5~8.5			5.5~6.5, 8.5~9	<5.5, >9
TDS	≤300	≤500	≤1000	≤2000	>2000
TH	≤150	≤300	≤450	≤550	>550
Cl^-	≤50	≤150	≤250	≤350	>350
SO_4^{2-}	≤50	≤150	≤250	≤350	>350
F^-	≤1.0			≤2.0	>2.0
NO_3^- (measured as N)	≤2.0	≤5.0	≤20	≤30	>30

Table 6.3 Classification table of EWQI results (Liu et al. 2022)

EWQI	Class	Water quality
<25	I	Water quality is excellent and can be used for various uses
25~50	II	Water quality is good and suitable for various uses
50~100	III	Water quality is medium and suitable for centralized domestic drinking water
100~150	IV	Water quality is poor and can be used after appropriate treatment
>150	V	Water quality is very poor and it is not suitable for drinking

Table 6.4 Overall water quality classification criteria based on EWQI

Water quality class	Water quality	Number of water samples	Percentages
I	Excellent	5	4.2%
II	Good	62	52.1%
III	Medium	39	32.8%
IV	Poor	10	8.4%
V	Very poor	3	2.5%

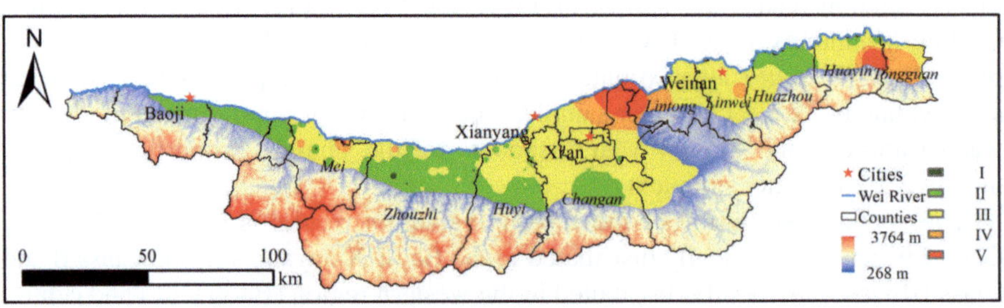

Fig. 6.9 Distribution map of EWQI calculation results

distributed around Xi'an and Weinan City. Class IV and Class V water are primarily concentrated in Mei County, Lintong District of Xi'an, and the eastern part of the northern Qinling Mountains (lower reaches of the Weihe River: eastern Huayin City and Tongguan District). Overall, the spatial distribution of groundwater quality in the northern Qinling Mountains presents a significant pattern where quality is better in the west than in the east and better in the south than in the north (water quality in the upper stream is better than in the downstream).

The reason for this phenomenon is that from the Qinling to the Wei River, along the groundwater flow path, the groundwater runoff speed slows down, leading to an increase in the residence time of dissolved ions. Meanwhile, the Guanzhong Plain is a major grain-producing area in Shaanxi Province, characterized by intense agricultural activities, rapid urbanization, low altitude, and strong evaporation. In summary, the deterioration of groundwater quality from the Qinling to the lower reaches of the Wei River is the result of the synergistic effect of increasing intensity of human activities and intensified natural evaporation and concentration.

6.5 Summary

This chapter presents a comprehensive investigation of the hydrochemical characteristics, controlling processes, and quality status of groundwater in the northern Qinling Mountains. The main conclusions are as follows:

1. Groundwater in the study area is weakly alkaline, with pH ranging from 6.83 to 9.12. The majority of samples are classified as freshwater based on total dissolved solids (TDS), although some show elevated salinity. Groundwater hardness is generally high, with many samples falling into the "hard" or "very hard" categories. Ca^{2+} and bicarbonate HCO_3^- are the dominant ions, and the principal hydrochemical type is HCO_3-Ca·Mg, with minor contributions from SO_4·Cl-Ca·Mg types. Notably, nitrate concentrations are elevated in agricultural regions, and fluoride concentrations exhibit high spatial variability.
2. Hydrochemical processes are primarily controlled by rock weathering, as indicated by Gibbs diagrams and supported by ionic ratio analysis. Carbonate dissolution (from calcite and dolomite) is a major contributor of Ca^{2+}, Mg^{2+}, and HCO_3^-, while halite and gypsum dissolution contribute Na^+, Cl^-, and SO_4^{2-}. Cation exchange also influences ion concentrations, especially for Na^+ and Ca^{2+}. Correlation analysis reveals strong relationships among TDS, major cations, and anions, supporting a common geogenic origin and the modifying effects of anthropogenic inputs in localized areas.
3. Based on EWQI, groundwater quality is generally favorable across the study area. Approximately 89% of samples fall within Class I to III, indicating excellent to moderate suitability for drinking. Only 10.9% of samples are classified as poor or very poor (Class IV and V), and these are sporadically distributed without forming extensive zones of deterioration. The spatial distribution of ground-

water quality in the northern Qinling Mountains presents a significant pattern where quality is better in the west than in the east and better in the south than in the north, indicating a better water quality in the upper stream than in the downstream. Nitrate contamination, largely associated with agricultural activities, poses localized health risks and warrants ongoing monitoring.

References

Argamasilla M, Barbera JA, Andreo B (2017) Factors controlling groundwater salinization and hydrogeochemical processes in coastal aquifers from southern Spain. Sci Total Environ 580:50–68. https://doi.org/10.1016/j.scitotenv.2016.11.173

Belkhiri L, Mouni L, Boudoukha A (2012) Geochemical evolution of groundwater in an alluvial aquifer: Case of El Eulma aquifer East Algeria. J Afr Earth Sci 66-67:46–55. https://doi.org/10.1016/j.jafrearsci.2012.03.001

Bradai A, Douaoui A, Bettahar N et al (2016) Improving the prediction accuracy of groundwater salinity mapping using indicator Kriging method. J Irrig Drain Eng 142:04016023. https://doi.org/10.1061/(ASCE)IR.1943-4774.0001019

Chen J, Gao Y, Qian H et al (2021) Hydrogeochemical evidence for fluoride behavior in groundwater and the associated risk to human health for a large irrigation plain in the Yellow River Basin. Sci Total Environ 800:149428. https://doi.org/10.1016/j.scitotenv.2021.149428

Farid I, Trabelsi R, Zouari K et al (2013) Geochemical and isotopic study of surface and groundwaters in Ain Bou Mourra basin, central Tunisia. Quat Int 303:210–227. https://doi.org/10.1016/j.quaint.2013.04.021

Gibbs RJ (1970) Mechanisms controlling world water chemistry. Science 170:1088–1090

Helstrup T, Jorgensen NO, Banoeng-Yakubo B (2007) Investigation of hydrochemical characteristics of groundwater from the Cretaceous-Eocene limestone aquifer in southern Ghana and southern Togo using hierarchical cluster analysis. Hydrogeol J 15:977–989. https://doi.org/10.1007/s10040-007-0165-1

Hoogland T, Heuvelink GBM, Knotters M (2010) Mapping water-table depths over time to assess desiccation of groundwater-dependent ecosystems in the Netherlands. Wetlands 30:137–147. https://doi.org/10.1007/s13157-009-0011-4

Jia H, Qian H, Zheng L et al (2020a) Alterations to groundwater chemistry due to modern water transfer for irrigation over decades. Sci Total Environ 717:137170. https://doi.org/10.1016/j.scitotenv.2020.137170

Jia H, Howard K, Qian H (2020b) Use of multiple isotopic and chemical tracers to identify sources of nitrate in shallow groundwaters along the northern slope of the Qinling Mountains, China. Appl Geochem 113:104512. https://doi.org/10.1016/j.apgeochem.2019.104512

Liu L, Wu J, He S, Wang L (2022) Occurrence and distribution of groundwater fluoride and manganese in the Weining plain (China) and their probabilistic health risk quantification. Expo Health 14(2):263–279. https://doi.org/10.1007/s12403-021-00434-4

Osiakwan GM, Appiah-Adjei EK, Kabo-Bah AT et al (2021) Assessment of groundwater quality and the controlling factors in coastal aquifers of Ghana: an integrated statistical, geostatistical and hydrogeochemical approach. J Afr Earth Sci 184:104371. https://doi.org/10.1016/j.jafrearsci.2021.104371

Piper AM (1944) A graphic procedure in the geochemical interpretation of water-analyses. EOS Trans Am Geophys Union 25:914–928. https://doi.org/10.1029/TR025i006p00914

Qian H, Ma Z, Li P (2012) Hydrogeochemistry. Geological Publishing House, Beijing. (In Chinese)

Schuster PF, Reddy MM, LaBaugh JW et al (2003) Characterization of lake water and ground water movement in the littoral zone of Williams Lake, a closed-basin lake in north central Minnesota. Hydrol Process 17:823–838. https://doi.org/10.1002/hyp.1211

Zhang Q, Qian H, Xu P et al (2021a) Effect of hydrogeological conditions on groundwater nitrate pollution and human health risk assessment of nitrate in Jiaokou Irrigation District. J Clean Prod 298:126783. https://doi.org/10.1016/j.jclepro.2021.126783

Zhang Q, Qian H, Xu P et al (2021b) Groundwater quality assessment using a new integrated-weight water quality index (IWQI) and driver analysis in the Jiaokou Irrigation District, China. Ecotoxicol Environ Saf 212:111992. https://doi.org/10.1016/j.ecoenv.2021.111992

Zhang C, Hou X, Li X et al (2021c) Hydrogeochemical characteristics and evolution mechanism of karst groundwater in the catchment area of the Sangu Spring. Hydrogeol Eng Geol 48(3):62–71. (In Chinese with English Abstract)

Chapter 7
Surface Water–Groundwater Interactions in Selected River Basins

Abstract The interaction between surface water and groundwater significantly impacts the spatial distribution and evolution of water resources and hydrochemical components. Studying the exchange between groundwater and surface water is of great significance for watershed water resource protection, riparian ecosystem conservation, and water pollution prevention. This chapter focuses on three representative river systems in the northern Qinling Mountains: the Shidi River (Weinan City), the Xin River (Xi'an City), and the Qingshui River (Baoji City), and surface water–groundwater interactions in these three river basins were investigated using connectivity index, hydrochemistry, isotopes, and statistical analysis.

Keywords Hydraulic connectivity · Isotopic method · End-member mixing model · River–groundwater interaction · Hydrochemical characteristics

7.1 Introduction

Surface runoff and subsurface runoff, as vital components of the natural hydrological cycle, often exhibit close hydraulic connections, interacting and transforming into each other spatially and temporally. The International Association of Hydrological Sciences (IAHS) and the International Association of Hydrogeologists (IAH) formally placed the topic of surface water–groundwater interactions on their conference agendas for discussion in 1986 and 1994, respectively (Tao 2012), marking this topic well researched among hydrogeological researchers.

A variety of methods are available for investigating interactions between surface water and groundwater, which can be broadly classified into hydrodynamic approaches and hydrochemical methods, including the use of isotopes. Hydrodynamic methods typically assess the direction and magnitude of water exchange by comparing water levels between surface water bodies and adjacent groundwater systems. These techniques are generally straightforward, conceptually intuitive, and capable of yielding accurate results. However, their application often incurs high costs, as they require boreholes that penetrate the aquifer, although

P. Li et al., *Water Resources Research in Northern Qinling Mountains, China*, Springer Hydrogeology, https://doi.org/10.1007/978-3-032-09395-0_7

existing wells are frequently utilized for such measurements. For example, Wang et al. (2021) examined the vertical infiltration rate of surface water into groundwater using long-term monitoring data from Baiyangdian Lake, offering valuable insights for ecological water replenishment strategies. Similarly, Shu et al. (2022) investigated the exchange between groundwater and surface water in the Suzhou section of the Xinbian River by analyzing hydraulic head differences across multiple cross-sections, applying Darcy's law to quantify the flow. In another study, Zhang (2022) focused on the Weinan section of the Wei River, measuring groundwater levels during the wet season and employing a multi-method approach to comprehensively assess the hydraulic connectivity between surface water and groundwater in the region.

Compared to hydrodynamic approaches, hydrochemical and isotopic methods are more widely applied in the study of surface water–groundwater interactions. The hydrochemical method typically involves selecting conservative tracers, such as chloride (Cl^-), and inferring water exchange by comparing the concentrations of these indicators in surface water and adjacent groundwater at the same location. Similar concentrations between the two water bodies suggest that mixing or exchange has occurred. However, a major limitation of this method is its inability to determine the direction of exchange, that is, whether groundwater is recharging surface water or vice versa. As a result, hydrochemical techniques are often used as supplementary tools or in combination with other methods to more accurately characterize exchange dynamics.

With the rapid advancement of isotopic techniques, stable isotopes of hydrogen and oxygen (δ^2H and $\delta^{18}O$) have become widely used for tracing surface water–groundwater interactions (Xu et al. 2023; Zhang et al. 2022a). These isotopes are valued for their natural abundance, conservative behavior, and distinct signatures in different water sources. Like hydrochemical tracers, stable isotopes can indicate mixing between surface and groundwater; however, when used in conjunction with end-member mixing models, they offer the additional advantage of quantifying the degree of exchange, such as estimating the proportion of surface water contributing to groundwater recharge or vice versa. For example, Zhang et al. (2020) analyzed the spatial distribution of δ^2H and $\delta^{18}O$ in river water and surrounding groundwater to reveal the exchange relationships in the western plain of the Ili River Valley. Navarro et al. (2017) applied isotope methods to investigate river recharge to groundwater under arid and semi-arid conditions in the Andarax River basin, southeastern Spain. Similarly, Mohammed et al. (2014) combined hydrodynamic, hydrochemical, and isotopic methods to explore the recharge sources of shallow groundwater and its interaction with the Allier River in Clermont-Ferrand, France.

In this chapter, connectivity index, hydrochemistry, and isotopes were conjunctively used to analyze the exchange relationship of three river systems in the northern Qinling Mountains. The recharge proportions using an end-member mixing model were also estimated, aiming to understand regional water resource cycling and provide references for the protection, development, and utilization of groundwater resources in the northern Qinling Mountains.

7.2 Methodology

7.2.1 *Selected River Systems*

The surface water in the northern foot of the Qinling Mountains belongs to the Wei River Basin, a first-level tributary of the Yellow River. In the Qinling Mountain area within this region, the mountains are high and the slopes are steep, with rivers and streams cutting deeply (Zhang et al. 2022b). There are numerous rivers and mountain passes, all characterized by narrow valleys and steep slopes. The rivers have short sources and rapid flows, with a low sediment content. There are as many as 108 major mountain valleys above the passes, and countless small ones. After the water from the mountain valleys flows out of the passes, they converge into 58 tributaries that enter the plain section. Eventually, 34 rivers are formed to flow into the Wei River, and 2 rivers flow into the Yellow River.

The length of the rivers in the northern foot of the Qinling Mountains is generally 10–100 km, and their drainage areas are usually 10–500 km^2. The northern foothills of the Qinling Mountains is rich in water resources with good water quality. The water resource quantity is approximately 3.815 billion m^3, accounting for 46% of the total water resources in the Guanzhong region (Li et al. 2025). It is an important water source for daily life and production in the Guanzhong area, meeting the water needs for daily life, production and ecology of about 12 million people in four large and medium-sized cities and a number of small cities in Guanzhong (Fig. 7.1).

1. Qingshui River

   Qingshui River is a first-level tributary on the southern bank of the Wei River. It originates from the northern foothills of the Qinling Mountains in Maying Town, Weibin District, Baoji City, and exits the mountain pass at Liaojiagou Village (the main factory area of Baotai Metal Products Co., Ltd.). The river section above the mountain pass flows through deep mountainous areas, with a narrow river channel, a large riverbed gradient, and a riverbed mainly composed of pebbles and block stones. The vegetation on both banks is relatively good, and there is only Yaojialing Village along the way with a small permanent population. The river empties into the Wei River at a location one kilometer west of the Wolongsi Wei River Bridge.

   From the mountain pass down to the confluence with the Wei River, the river channel widens, the riverbed gradient gradually decreases, and the riverbed material becomes finer, mainly consisting of pebbles and fine sand. The catchment area of Qingshui River is 162.5 km^2, with a total length of 28.4 km, an average gradient of 3.52‰, and an average annual runoff of 100.408 million cubic meters over the years.

   The Qingshui River Basin is located within Baoji City, and the river mainly flows through Weibin District. A large number of titanium processing plants and villages are distributed on both banks of the river, with a dense population and numerous sources of pollution from production and daily life, resulting in poor

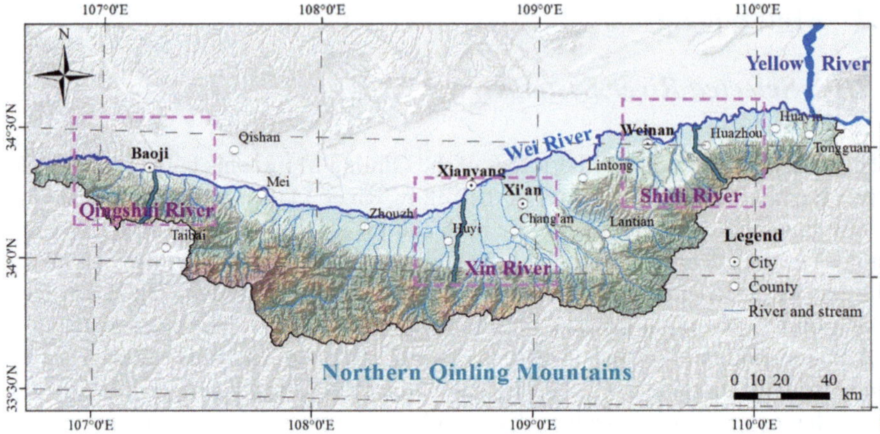

Fig. 7.1 Location map of typical rivers in the northern Qinling Mountains (Lower Reaches of the Wei River: Shidi River; Middle Reaches of the Wei River: Xin River; Upper Reaches of the Wei River: Qingshui River)

water quality of the river. Qingshui River originates from Gaojiaya, and along its course, there are Gaojiaya Canal, Shanmenkou Canal, Wenquan Canal, and Guojiacun Canal. After the founding of the People's Republic of China, Qingshui River was renovated, with 6 reservoirs built and over 20,000 mu of farmland irrigated.

2. Xin River

The Xin River is a first-level tributary on the southern bank of the River. It originates from the shallow mountainous area at the northern foot of the Qinling Mountains, flows from south to north through towns and townships such as Shijing, Yuxia, Wuzhu, Cangyou, and Dawang in Huyi District, Xi'an. It enters Fengxi New City at Dongjiangdu Village, Diaotai Town, Fengxi New City, then passes through Caojiatan Village and Wangdao Village, and finally converges into the Wei River at Majiazhai Village. With a catchment area of 303.8 km^2, the river has a total length of 40 km.

The Xin River is an artificial river. It was artificially excavated due to frequent floods caused by low-lying terrain at the border between Chang'an and Huyi District, hence the name "Xin" (meaning "New River"). Therefore, it is essentially a drainage channel, and its water regime changes with rainfall. During the flood season, the river's water volume is mainly replenished by rainfall, while in the dry season, it relies primarily on stable groundwater recharge.

The Xin River Basin is mainly located in Xi'an, with the river flowing primarily through Huyi District. Due to the limited scope of comprehensive regulation works on the Xin River, which are mainly concentrated in key urban sections, most of the rural embankments in other sections were built in the 1960s and 1970s. These embankments have low flood control standards, and after years of operation, they suffer from severe bank slope erosion and river channel siltation, resulting in significant flood control pressure. In addition, there are numerous towns, villages, and industrial parks along the river. The phenomena of unauthorized drainage, random dumping, and illegal cultivation in the river channel are relatively serious, making the problems of the aquatic ecological environment increasingly prominent.

3. Shidi River

   The Shidi River is a first-level tributary on the southern bank of the Wei River. It originates from Shibanchagou on the northern slope of Laoniu Mountain (with an elevation of 2539.6 m), the main ridge of the Qinling Mountains, and is situated in Huazhou District, Weinan City (Li et al. 2016). The source elevation ranges from 2000 to 2640 m, and after flowing out of the mountain pass, it drops to 500–600 m. With steep slopes and a short course, the river traverses the piedmont alluvial fan apron, where the elevation further decreases to 340–350 m, before flowing gently into the Wei River. The drainage area of the river basin is 134.8 km^2, the total length of the river is 36.8 km, and the total drop reaches 1439 m.

   Floods in the Shidi River are mostly triggered by heavy rains. Owing to the high mountains and steep slopes, water converges rapidly and the flow path is short, resulting in sudden rises and falls of floods, which typically last for approximately 6–7 h. According to surveys, the water in the upper reaches of the Shidi River is clear. The terrain in this area is high in the west and low in the east, surrounded by mountains with lush vegetation. There are numerous gravel accumulations on both banks, accompanied by near-shore sediment deposition. The rocks in the river channel exhibit poor sorting, and there is little garbage in the surrounding environment.

   The river dries up at the exit of the mountain pass, with a few weeds growing in the wide river channel. The dry period is not long, and the riverbed is covered with a large quantity of high-quality fine sand, with few large stones. Local residents often come here for sand extraction. This also confirms that the Shidi River currently possesses the characteristics of a seasonal river, and its protection should be reinforced.

7.2.2 Connectivity Index (CI) Method

The connectivity index (CI) is a simple method for assessing the connectivity between surface water and groundwater, primarily considering four factors: groundwater level, channel sediment, geology, and geomorphology (Oyarzu'n et al. 2014; Li et al. 2016). Each parameter is scored based on specific conditions, and the CI is calculated using Eq. (7.1):

$$\mathrm{CI} = 3\mathrm{D} + 5\mathrm{S} + 5\mathrm{G} + 2\mathrm{GM} \tag{7.1}$$

Table 7.1 Indicator scoring for CI calculation

Indicators	Classification	Scores
Water table depth (D)	<10 m	5
	10–20 m	3
	>20 m	0.5
Channel sediment (S)	Sand, gravel	5
	Silty sand, silt	3
	Silt, clay loam	−1
	Clay	−4
Geology (G)	Gravel, sand	5
	Clay, sand	3
	Clay	−4
Geomorphology (GM)	Erosional environment	5
	Depositional environment	1
	Terrace	0

Where, CI is the connectivity index; D is the groundwater level indicator score; S is the channel sediment indicator score; G is the geological indicator score, and GM is the geomorphological indicator score.

In this study, groundwater levels in shallow wells along the rivers were measured, making them suitable for river-aquifer connectivity analysis. The channel sediment indicator was determined through field surveys and laboratory particle size analysis. Geological indicator information (lithology) came from field observations and pre-existing geological maps, while geomorphological characteristics were derived from field investigation and data collection.

The classification criteria for CI are: $-78.5 < CI < -24$ indicates low connectivity; $-24 \leq CI \leq 53$ indicates medium connectivity; $53 < CI \leq 75$ indicates high connectivity (Oyarzu'n et al. 2014). In this study, the CI method was primarily used to determine the potential of surface water and groundwater interaction in the Shidi River, Xin River and Qingshui River basins. The scores for groundwater level, channel sediment, geology, and geomorphology are summed according to Eq. (7.1) to obtain the CI. A higher CI value indicates greater river-aquifer connectivity (Table 7.1).

7.2.3 Hydrochemical Method

The hydrochemical method aids in understanding the water cycle, water environment changes, and interconnections between different water bodies. Analyzing the hydrochemical connection between groundwater and surface water requires not only looking at differences in ion concentrations between water types but also considering whether the proportional composition of ions is consistent (Yang et al. 2025). When both the absolute and relative concentrations of all ions are similar in different water bodies, their hydrochemical connection is considered strong. Piper trilinear diagrams and Durov diagrams can visually display the correlation in hydrochemical composition between surface water and groundwater, and identify differences between groundwater and surface water along the river. These diagrams are primarily plotted using 10 indicators: pH, TDS, K^+, Na^+, Ca^{2+}, Mg^{2+}, Cl^-, SO_4^{2-}, HCO_3^-, and CO_3^{2-}. The connectivity and exchange relationship can be determined based on the ion proportions.

The exchange between surface water and groundwater often involves material and energy transfer. The hydrochemical characteristics of different water bodies can reveal the circulation processes of atmospheric precipitation, surface water, and groundwater, as well as interactions between them (Li et al. 2018). The connectivity between different water bodies can be determined using the concentration difference method (Eq. 7.2):

$$\frac{|C_1 - C_2|}{\frac{1}{2}(C_1 + C_2)} \times 100\% \tag{7.2}$$

Where, C_1 is the concentration of an ion in the upper aquifer groundwater; C_2 is the concentration of the same ion in the lower aquifer groundwater. If the concentration difference for an ion is less than 20%, connectivity between the two adjacent aquifers is high; if between 20 and 40%, connectivity is medium; if exceeding 40%, connectivity is low.

In this chapter, Cl^- was primarily used to calculate the ion concentration difference between surface water and groundwater in the Shidi River, Xin River, and Qingshui River, thereby determining their degree of connectivity (Eq. 7.3):

$$\frac{|Cl_1 - Cl_2|}{\frac{1}{2}(Cl_1 - Cl_2)} \times 100\% \tag{7.3}$$

7.2.4 *Isotopic Discrimination Method*

Stable hydrogen and oxygen isotopes (δ^2H and $\delta^{18}O$) are widely recognized as effective natural tracers for studying the sources, flow paths, and interactions of surface water and groundwater (West et al. 2014). These isotopes are sensitive to climatic and geographic variables, typically exhibiting temperature, altitude, and latitude effects. For example, in regions where surface water and groundwater are closely connected, their hydrochemical and isotopic signatures often show strong similarity. In contrast, weak hydraulic connectivity is usually reflected by significant differences in isotopic composition.

Because of their sensitivity to environmental conditions, δ^2H and $\delta^{18}O$ have been extensively used to trace the origins and evolution of water bodies, and to assess surface water–groundwater exchange processes. One of the foundational tools in isotope hydrology is the Global Meteoric Water Line (GMWL), derived by Craig through regression analysis of data from the global precipitation isotope network (Eq. 7.4) (Terzer-Wassmuth et al. 2023). The GMWL provides a reference framework for interpreting isotopic data and understanding water source signatures under various climatic regimes.

$$\delta^2\text{H} = 8\delta^{18}\text{O} + 10 \tag{7.4}$$

However, the slope and intercept of GMWL often differ from those of the Local Meteoric Water Line (LMWL) due to regional variations in temperature, humidity, and precipitation patterns. Therefore, for hydrological studies at the basin or watershed scale, it is more appropriate to use the LMWL, which better represents the local isotopic characteristics of precipitation.

By plotting both the GMWL and LMWL alongside the measured isotopic composition of water samples (δ^2H and δ^{18}O), the origin and recharge source of groundwater can be inferred based on the relative position of the samples to these meteoric water lines. Water that plots near the LMWL generally indicates recent meteoric recharge, whereas deviations from the line may suggest evaporation effects, mixing processes, or water–rock interactions (Ren et al. 2024).

Stable hydrogen and oxygen isotope ratios are commonly expressed as per mille (‰) deviations relative to Vienna Standard Mean Ocean Water (VSMOW), using the delta (δ) notation, as defined by the following equation (Eq. 7.5) (Zhao et al. 2021):

$$\delta\left(‰\right) = \frac{R_{\text{sample}} - R_{\text{standard}}}{R_{\text{standard}}} \times 1000 \tag{7.5}$$

Where, R_{sample} is the ^{18}O/^{16}O or ^{2}H/H ratio in the surface water or groundwater sample; R_{standard} is the ^{18}O/^{16}O or ^{2}H/H ratio in VSMOW.

In systems where surface water and groundwater are hydraulically connected, the degree of connectivity can be inferred from their stable isotopic compositions. Strong hydraulic connectivity is typically indicated by similar δ^2H and δ^{18}O values, while poor connectivity is reflected by significant isotopic differences between the two water bodies. Similar to hydrochemical assessments, threshold criteria can be applied to isotopic data to evaluate connectivity. Specifically, 20% and 40% difference thresholds are often used to classify the degree of exchange. Given that stable isotopes are expressed as per mille (‰) deviations from the VSMOW, these thresholds correspond to specific isotopic ranges: for δ^{18}O, the thresholds are approximately ±0.7‰ to ±1.4‰, and for δ^2H, approximately ±4‰ to ±9‰. These ranges provide a practical basis for determining whether surface water and groundwater are well-connected, partially connected, or largely disconnected.

7.2.5 Cluster Analysis

Cluster analysis is a multivariate classification technique that groups samples or variables based on the similarity of their characteristics (Bu et al. 2020). It is generally categorized into two types: Q-mode clustering, which focuses on grouping sample sites to explore spatial relationships, and R-mode clustering, which classifies variables or parameters based on their correlations. This method is widely applied across disciplines such as hydrology, geology, meteorology, mining, industrial process analysis, and environmental science.

Various clustering algorithms are available, including the farthest neighbor, nearest neighbor, centroid, and Ward's method. Among these, Ward's method, also known as the minimum variance method, has demonstrated strong performance in hydrological and water resources studies and was therefore selected in this study to identify clusters relevant to surface water–groundwater interactions. The Euclidean distance was used as the dissimilarity metric, and all data were standardized using Z-score transformation prior to analysis to eliminate the effects of differing units and scales.

Ward's method operates by minimizing the total within-cluster variance at each step of the hierarchical clustering process. Initially, each observation is treated as an individual cluster. At each iteration, the algorithm computes the increase in the total within-cluster sum of squares that would result from merging each possible pair of clusters, and merges the pair that leads to the smallest increase. This process continues until all observations are merged into a single cluster. The specific steps of the algorithm are detailed below:

For the classification of n regional samples, denoted as G_1, G_2, …, G_t, let $X_i^{(t)}$ denote i_{th} sample in cluster G_t. $X_i^{(t)}$ is a p-dimensional vector, corresponding to the p clustering indices or variables). Let n_t denote the number of samples in cluster G_t, and let $\overline{X}^{(t)}$ be the centroid (mean vector) of G_t, calculated as the average of all sample vectors within the cluster. The sum of squared deviations (S_t) within G_t is then defined as Eq. (7.6):

$$S_t = \sum_{i=1}^{n_t} \left(X_t^{(t)} - \overline{X}^{(t)}\right)' \left(X_i^{(t)} - \overline{X}^{(t)}\right) \tag{7.6}$$

The total sum of the squared deviations within-cluster (SSD) for k clusters is (Eq. 7.7):

$$SSD = \sum_{t=1}^{k} S_t = \sum_{t=1}^{k} \sum\nolimits_{i=1}^{n_t} \left(X_t^{(t)} - \overline{X}^{(t)}\right)' \left(X_i^{(t)} - \overline{X}^{(t)}\right) \tag{7.7}$$

Euclidean distance refers to the true distance between two points in n-dimensional space, which can be calculated using Eq. (7.8):

$$\rho(A,B) = \sqrt{\sum_{i=1}^{n} \left[\mu_A(x_i) - \mu_B(x_i)\right]^2} \tag{7.8}$$

Where, $\rho(A, B)$ is the Euclidean distance between points A and B in n-dimensional space; $\mu_A(x_i)$ is the coordinate of vector A on the i_{th} dimension; $\mu_B(x_i)$ is the coordinate of vector B on the i_{th} dimension. A smaller Euclidean distance indicates closer vectors. Ward's clustering method stems from analysis of variance and enables comprehensive classification based on multiple characteristics of the objects studied. It minimizes the increment of *SSD*, effectively highlighting homogeneity within clusters and heterogeneity between clusters, fully reflecting the comprehensive differences between study objects.

7.2.6 Sample Collection and Analysis

To study the groundwater–surface water exchange in the northern foot of the Qinling Mountains, fieldwork was carried out targeting the Shidi River, Xin River, and Qingshui River. The main tasks included: collection of regular surface water and groundwater samples, collection of isotope samples and on-site determination of water sample indicators.

In the Shidi River, a total of ten samples were collected, including three surface water samples, six groundwater samples, and one rainwater sample. Stable hydrogen and oxygen isotope testing was conducted simultaneously on these ten water samples. In addition, seven groundwater levels were measured. In the Xin River, a total of 15 groundwater levels were measured. In the Qingshui River, a total of five water samples were collected, including three surface water samples and two groundwater samples.

7.3 Surface Water–Groundwater Exchange Relationship in the Shidi River

The distribution of sampling sites within the Shidi River Basin is illustrated in Fig. 7.2. The Shidi River originates in the Qinling Mountains and flows northward through a piedmont alluvial plain, traverses the terraces of the Wei River, and eventually merges with the Wei River. The floodplain and terraces of the Wei River are primarily composed of sandy soils and fine sand, with the first terrace extensively covered by a mixture of sand, gravel, and silty clay (Lu et al. 2025). The geological structure of the study area is dominated by interbedded layers of coarse sand, sandy clay, and fine sand, which provide a permeable medium conducive to surface water–groundwater interaction.

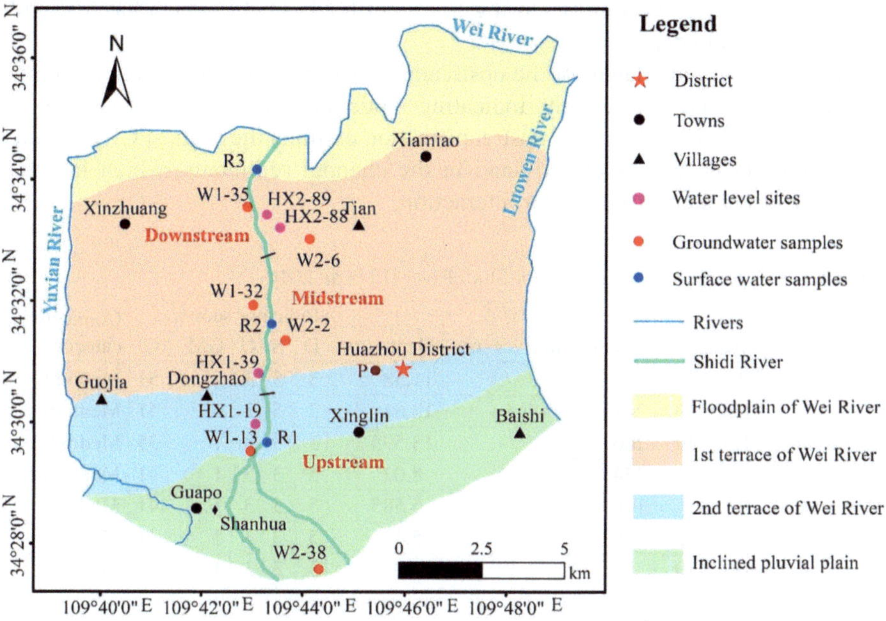

Fig. 7.2 Sampling sites of the study area

A total of ten water samples were collected for analysis, comprising three surface water samples (R1, R2, R3) and six groundwater samples (W1-13, W1-32, W1-35, W2-2, W2-6, W2-38) and one rain water sample (P). The groundwater samples were obtained from shallow wells (depth <30 m) that are commonly used for domestic and agricultural purposes by the local population. Surface water samples were collected from the upstream, midstream, and downstream sections of the Shidi River to capture longitudinal variations in water chemistry and to support the analysis of surface water–groundwater interactions.

7.3.1 Hydraulic Connection Using CI Method

The Connectivity Index (CI) results for various sections of the Shidi River are summarized in Table 7.2. The upstream section of the river is characterized by coarse channel sediments, predominantly gravel and sand, which yield the highest sediment score of 5, reflecting high permeability and infiltration potential. In contrast, midstream and downstream sections are composed mainly of sandy loam or silty loam, resulting in lower sediment scores of 3 due to reduced hydraulic conductivity.

Groundwater depth exhibits a decreasing trend from upstream to downstream, which influences the groundwater level scores. Upstream wells (e.g., W1-13, HX1-19) are generally deeper (>10 m), corresponding to a moderate groundwater level score of 3. Conversely, wells in the midstream and downstream sections are typically shallower, with most scoring 5, except for HX1-39, which has a slightly deeper water table depth.

The CI values calculated for the upstream, midstream, and downstream sections are 51, 35, and 41, respectively, indicating moderate connectivity throughout the basin (Table 7.2; Fig. 7.3). The upstream section exhibits the highest CI, attributed to the dominance of gravel and sand in the channel sediments, which facilitate strong surface water–groundwater interaction.

Table 7.2 Connectivity index results for Shidi River (Li et al. 2016)

Reaches	Wells	Distance to river (m)	Depth (m)	Indicator scores				CI	Connectivity category
				D	S	G	GM		
Upstream	W1-13	201	11.58	3	5	3	1	51	Medium
	HX1-19	247	13.61	3	5	3	1	51	Medium
Midstream	HX1-39	266	13.595	3	3	3	1	35	Medium
	W2-2	354	8.02	5	3	3	1	41	Medium
Downstream	W1-35	160	5.585	5	3	3	1	41	Medium
	HX2-88	460	5.43	5	3	3	1	41	Medium
	HX2-89	91	6.031	5	3	3	1	41	Medium

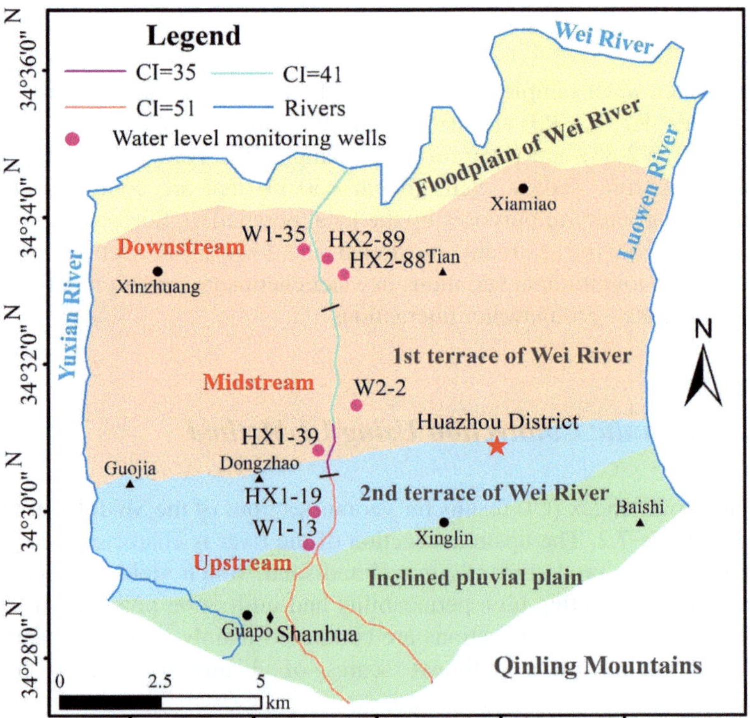

Fig. 7.3 Spatial variation of connectivity based on CI

The midstream section includes two subsections. In the southern subsection (HX1-39), the presence of sandy loam and silty loam channel sediments, coupled with a relatively deeper groundwater table, results in a lower CI of 35. In contrast, the northern subsection (W2-2), although sharing similar sediment composition, has a shallower water table, which increases its CI to 41.

The downstream section exhibits a CI comparable to the northern part of midstream section (CI = 41). Although sediment grain size decreases along the river's course, leading to reduced infiltration capacity, the shallower groundwater depths in the downstream region compensate for this by contributing to a higher groundwater level score, and thus a higher overall CI than the southern subsection.

Across all sections, the geological and geomorphological characteristics remain relatively consistent, indicating that their influence on surface water–groundwater connectivity is stable throughout the basin.

7.3.2 Hydraulic Connection Using Hydrochemistry and Cluster Analysis

To assess the hydraulic connection between surface water and groundwater in the Shidi River Basin, Durov diagrams were employed to analyze variations in major ion compositions (Fig. 7.4). The Durov diagram is a comprehensive hydrochemical tool that integrates triangular fields for major cations and anions, as well as a central square field representing overall hydrochemical types. It can also display pH and TDS, offering a multidimensional view of water chemistry.

In this study, the Durov plot revealed two distinct clusters of water samples. Cluster 1 comprises all groundwater samples, including those from the upstream (W1-13), midstream (W1-32 and W2-2), and downstream (W1-35 and W2-6) sections. Cluster 2 includes all surface water samples, namely R1 (upstream), R2 (midstream), and R3 (downstream).

The clear separation of these two clusters within the triangular fields of the Durov diagram indicates significant differences in major ion concentrations between surface water and groundwater. This distinct hydrochemical differentiation suggests a limited hydraulic connection between the two water bodies in the study area. The lack of convergence in hydrochemical facies implies that mixing or interaction between surface water and groundwater is minimal or highly localized.

Given that ions such as K^+, Na^+, Ca^{2+}, Mg^{2+}, SO_4^{2-}, and HCO_3^- are prone to adsorption, ion exchange, and other geochemical reactions, the more conservative ion chloride (Cl^-) was selected to assess surface water–groundwater connectivity. The concentration difference of Cl^- between surface water and groundwater was calculated using Eq. (7.2).

As shown in Table 7.3, the Cl^- concentration differences for the upstream, midstream, and downstream sections were 25.0%, 24.25%, and 20.5%, respectively. All values fall within the 20–40% threshold range, which is commonly interpreted as indicative of medium hydraulic connectivity.

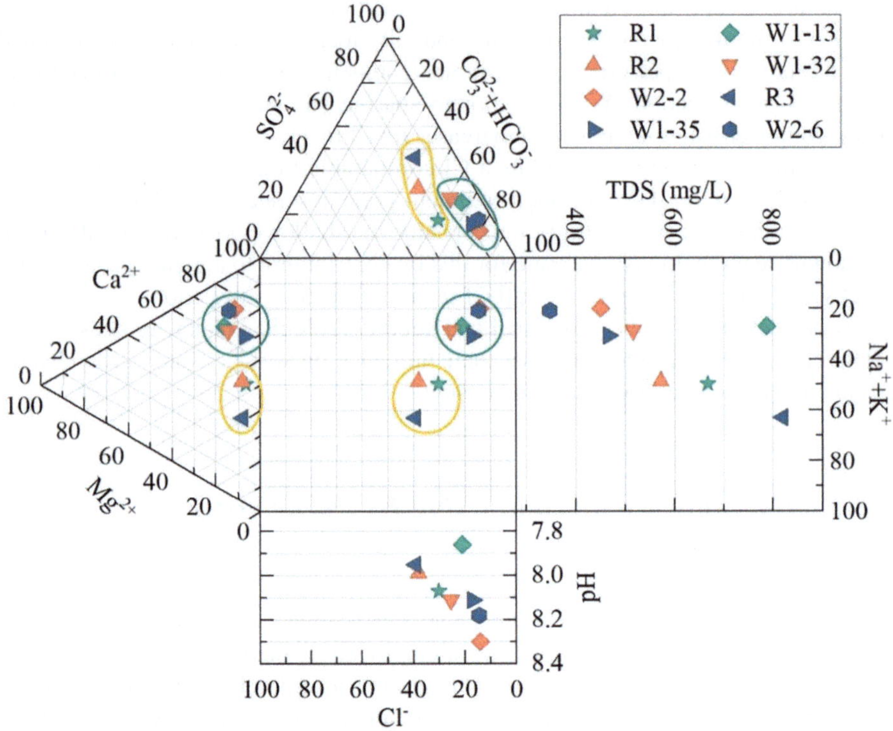

Fig. 7.4 Durov diagram of water samples

Table 7.3 The difference of Cl^- between groundwater and surface water in the Shidi River

Sections	Mean Cl^- of groundwater (mg/L)	Mean Cl^- of river water (mg/L)	Concentration difference (%)	Connectivity
Upstream	35	15	25	Medium
Midstream	32	16.5	24.25	Medium
Downstream	26	15	20.5	Medium

When combined with the results of the Durov diagram analysis, which also indicated a degree of separation yet not complete disconnection between surface and groundwater, the overall assessment suggests that the hydraulic connection between surface water and groundwater in the Shidi River Basin can be classified as medium.

To investigate the relationship between ion concentrations and assess water exchange, Hierarchical Cluster Analysis (HCA) was applied to both absolute and relative concentrations of major ions in surface water and groundwater samples. A Q-mode HCA was conducted using a comprehensive dataset of hydrochemical parameters including pH, TDS, TH, F^-, K^+, Na^+, Ca^{2+}, Mg^{2+}, Cl^-, SO_4^{2-}, HCO_3^-, COD, CO_2, NO_3^-, along with stable isotopes (δ^2H and $\delta^{18}O$). The resulting dendrogram is presented in Fig. 7.5.

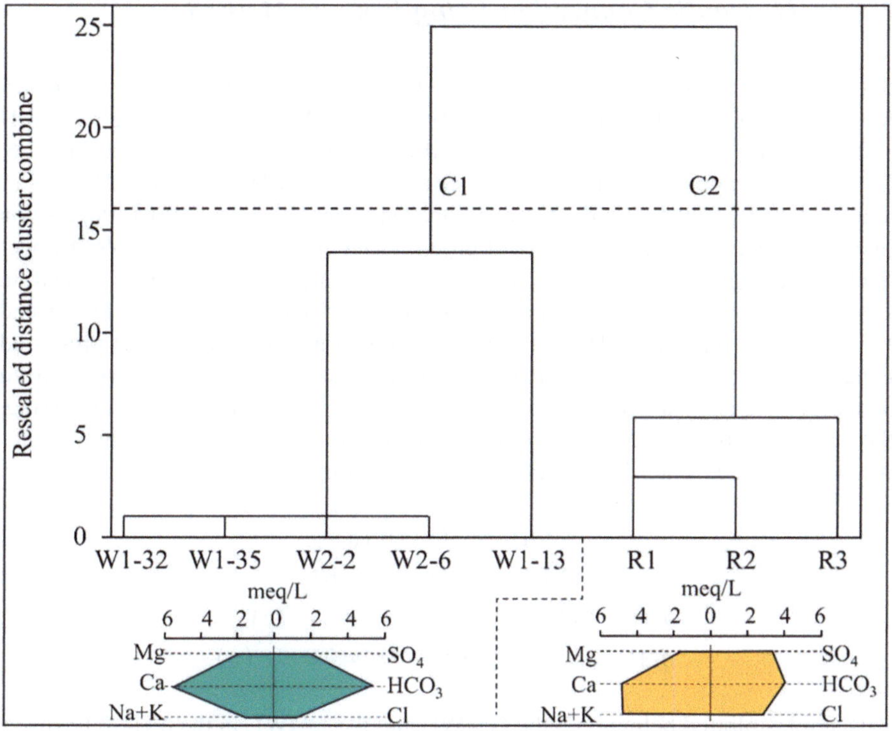

Fig. 7.5 Clusters obtained from HCA and the Stiff diagrams based on the mean concentrations of major ions in each cluster

The analysis revealed two distinct clusters: Cluster 1 (C1) comprises all groundwater samples, while Cluster 2 (C2) includes all surface water (river) samples. This clear separation between clusters reflects significant differences in chemical and isotopic characteristics, suggesting low hydraulic connectivity between surface water and groundwater in the study area.

Further insights are provided by the Stiff diagrams, which illustrate the average milliequivalent concentrations of major ions for each water type. In surface water, the dominant cations are Ca^{2+} (4.71 meq/L) and Na^{+} (4.13 meq/L), while the major anions are HCO_3^- (4.07 meq/L), followed by SO_4^{2-} (3.35 meq/L) and Cl^- (2.85 meq/L). In contrast, groundwater samples are primarily characterized by Ca^{2+} (5.70 meq/L) as the dominant cation and HCO_3^- (5.34 meq/L) as the major anion.

The differences in hydrochemical composition, as visualized in both the HCA dendrogram and the Stiff diagrams, further support the conclusion that the hydraulic connection between groundwater and surface water is limited in the Shidi River Basin.

7.3.3 Hydraulic Connection Using Isotopic Method

Stable isotopes were employed to evaluate the exchange relationship between surface water and groundwater in the Shidi River Basin. The relationship between δ^2H and δ^{18}O values for river water and groundwater samples is illustrated in Fig. 7.6. The analysis was conducted with reference to both the GMWL, shown as Eq. 7.9(a), and the LMWL for the Shidi River region, represented by Eq. 7.9(b). Regression lines were fitted to the isotopic data of groundwater and river water samples respectively. The regression equations for these two water sources are presented in Eq. 7.9(c, d), with high correlation coefficients of $r^2 = 0.984$ for groundwater and $r^2 = 0.999$ for river water. These strong linear relationships indicate well-preserved meteoric signatures and suggest that, while both water bodies are derived from meteoric sources, their distinct regression lines reflect limited isotopic mixing, thereby supporting the conclusion that the hydraulic connectivity between groundwater and surface water in the study area is not strong.

$$\begin{aligned} \delta^2\text{H} &= 8^{18}\text{O} + 10 \ (\text{a}) \\ \delta^2\text{H} &= 7.5^{18}\text{O} + 6.1 \quad (\text{b}) \\ \delta^2\text{H} &= 7.11^{18}\text{O} + 4.31 \quad r^2 = 0.984 \quad (\text{c}) \\ \delta^2\text{H} &= 3.92^{18}\text{O} - 26.52 \quad r^2 = 0.999 \quad (\text{d}) \end{aligned} \tag{7.9}$$

All samples plotted below both the GMWL and the LMWL in the δ^2H–δ^{18}O space, with regression lines exhibiting slopes lower than those of the meteoric lines. This pattern indicates δ^{18}O enrichment relative to meteoric water, a signature typically associated with evaporation effects. Notably, as one moves downstream along the flow direction, the distance of both groundwater and surface water samples from the

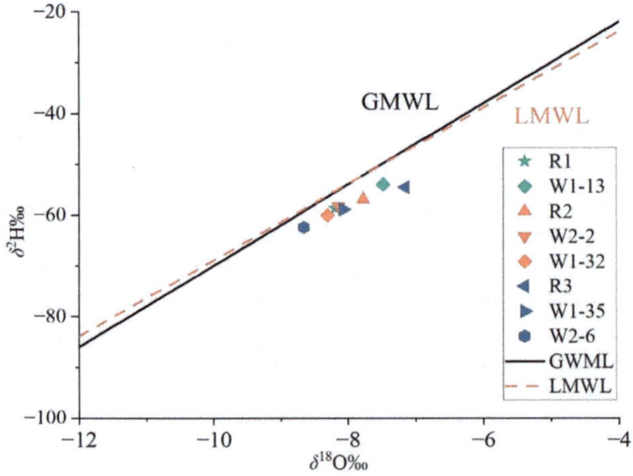

Fig. 7.6 Isotopic relationship diagram of groundwater and river water

meteoric lines increases, suggesting progressive $\delta^{18}O$ enrichment and thus intensified evaporation. Interestingly, the regression slope for river water is lower than that of groundwater, implying that evaporation effects are more pronounced in surface water than in groundwater.

Furthermore, the groundwater regression line closely parallels and lies near the LMWL, indicating that local precipitation is the dominant source of groundwater recharge. The relatively minor deviation from the meteoric line also suggests limited isotopic mixing with surface water, reinforcing earlier conclusions regarding weak hydraulic connectivity between groundwater and river water in the Shidi River Basin.

During the sampling campaign, groundwater $\delta^{18}O$ values ranged from −8.54‰ to −7.48‰, and $\delta^{2}H$ values ranged from −62.42‰ to −54.02‰. In comparison, river water $\delta^{18}O$ ranged from −8.19‰ to −7.15‰, and $\delta^{2}H$ from −58.67‰ to −54.60‰. The estimated connectivity assessment based on $\delta^{18}O$ and $\delta^{2}H$ values is summarized in Table 7.4. Notably, there is an overlap between the isotopic composition of river sample R1 and several groundwater samples, indicating a hydrological connection. This overlap suggests that the river section represented by R1 serves as an important recharge source for the local groundwater system, likely due to favorable geomorphological and hydrodynamic conditions in that reach of the river.

Statistical analysis of hydrochemical data revealed certain similarities between river water and adjacent groundwater, indicating potential interaction. The CI method classified all sections of the Shidi River as having medium connectivity. This method integrates multiple factors, including groundwater level, channel sediment properties, geological structure, and geomorphology. As geology and geomorphology were consistent across the study area, the observed spatial variation in CI values was primarily attributed to differences in groundwater depth and channel sediment characteristics.

Overall, multiple independent methods, including hydrochemical analysis, CI evaluation, stable isotope analysis, and hierarchical cluster analysis, converged on a consistent conclusion: the hydraulic connection between surface water and groundwater in the Shidi River Basin is of medium intensity. This consistency underscores the validity of using CI, hydrochemical, and isotopic approaches for evaluating hydraulic connectivity and groundwater recharge processes.

Each method offers unique advantages. The CI method, based on a straightforward formula and easily obtainable field parameters, yields efficient preliminary assessments without requiring extensive laboratory analyses, thereby reducing time and cost. In contrast, isotopic methods offer higher precision and deeper insight into recharge processes and flow dynamics but involve greater expense, more complex

Table 7.4 Connectivity between river water and groundwater estimated by different methods

Sections	CI	Hydrochemistry method	$\delta^{18}O$	$\delta^{2}H$	Final connectivity
Upstream	Medium	Medium	Medium	Medium	Medium
Midstream	Medium	Medium	Medium	Medium	Medium
Downstream	Medium	Medium	Medium	Medium	Medium

sampling protocols, and limited sample coverage, which can affect the scale and accuracy of assessments.

Therefore, when selecting methods to assess surface water–groundwater connectivity, it is important to balance cost, precision, data availability, and operational feasibility. An integrated approach that combines multiple techniques can enhance reliability while accommodating the constraints of individual study areas.

7.3.4 Estimation of Groundwater Recharge Contributions

Since surface water infiltration plays a significant role in recharging groundwater in the Shidi River Basin, end-member mixing analysis (EMMA) was employed to quantify the contribution rates of surface water infiltration (Hooper et al. 1990). Three recharge sources, i.e., precipitation, surface water infiltration, and lateral groundwater flow, were considered in the calculation, as they are the most important recharge sources to the groundwater system in this area.

7.3.4.1 EMMA Model

EMMA is a quantitative method used to interpret the hydrochemical signature of a sample as a mixture of different sources (end-members). In the study area, the primary sources of groundwater recharge during the dry season are precipitation, lateral inflow, and river water infiltration (Christophersen et al. 1990; Hooper et al. 1990). Therefore, a three-end-member EMMA was applied. The assumptions underlying EMMA are: (1) tracer concentrations in each end-member remain constant; (2) mixing between end-members is linear; and (3) tracers behave conservatively without undergoing chemical or physical alteration. In this chapter, $\delta^{18}O$, δ^2H, and Cl^- concentrations are used as tracers to estimate the relative contributions of each source. Based on above assumption, the EMMA can be expressed using Eq. (7.10) (Liu and Yamanaka 2012):

$$
\begin{aligned}
R_r &= \frac{(\delta_s-\delta_i)(c_p-c_i)-(c_s-c_i)(\delta_p-\delta_i)}{(\delta_r-\delta_i)(c_p-c_i)-(c_r-c_i)(\delta_p-\delta_i)} \\
R_P &= \frac{(\delta_s-\delta_i)(c_r-c_i)-(c_s-c_i)(\delta_r-\delta_i)}{(\delta_p-\delta_i)(c_r-c_i)-(c_p-c_i)(\delta_r-\delta_i)} \\
R_i &= 1-R_r-R_P
\end{aligned}
\tag{7.10}
$$

Where, R is the contribution rate, δ represents ^{18}O and ^2H values, c is Cl^- concentration. Subscripts s, i, p, r represent the sampled water, lateral inflow groundwater, precipitation, and river water infiltration, respectively. When samples fall outside the model domain defined by the three end-members, geometric methods are employed to estimate the contributions of these outliers. For this analysis, lateral inflow groundwater (Groundwater sample W2-38), local precipitation (Rain sample P), and upstream river water (River water R1) were selected as the end-members.

7.3.4.2 Contribution Rate Estimation

Groundwater recharge contributions were quantified using Eq. (7.10). The results of the EMMA (End-Member Mixing Analysis) model, based on the relationships between Cl^- and $\delta^{18}O$ as well as Cl^- and δ^2H, are presented in Fig. 7.7. Most groundwater and river water samples fall within the triangular mixing domain defined by three end-members: local precipitation (P), upstream groundwater (W2-38, representing lateral inflow), and river water (R1). This distribution suggests that groundwater recharge along the Shidi River is primarily derived from these three sources. However, sample R3 falls outside the EMMA mixing triangle (Fig. 7.7a), indicating that its source contributions could not be reliably estimated using the three-end-member EMMA model. Therefore, geometric methods were employed to estimate the recharge contributions for this sample.

Surface water sample R3 indicates two primary sources of recharge: local precipitation, contributing 26.3–40.5%, and upstream river water, contributing 59.5–69.6%. All groundwater samples are plotted within the ternary mixing diagram (Fig. 7.8), indicating contributions from all three sources. Estimated recharge contributions in the upstream section are as follows: precipitation 35.5–44.4%, river seepage 26.6–27.0%, and lateral groundwater inflow 28.9–35.5%. In the midstream section, the respective contributions are 13–17.8% from precipitation, 18.4–34.63% from river seepage, and 47.6–68.3% from lateral inflow. Downstream, the contributions shift to 5.4–22.4% from precipitation, 4.72–26.6% from river seepage, and 51.3–89.8% from lateral inflow.

The contribution of river seepage tends to decrease downstream, likely due to the progressive fining of channel sediments that limit infiltration. In the Shidi River

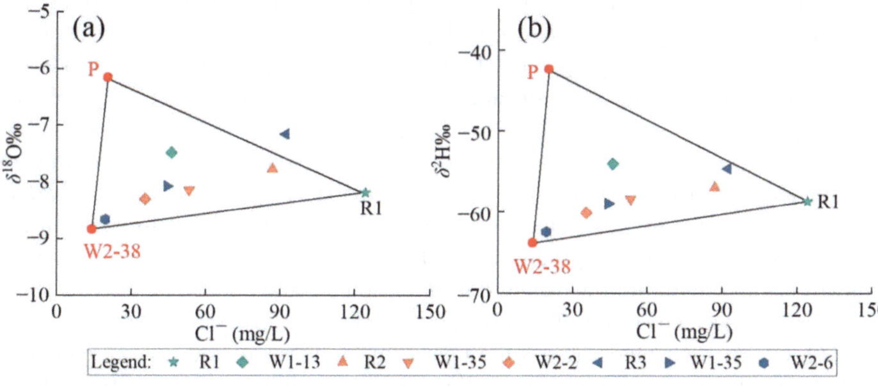

Fig. 7.7 End member mixing models established based on the relationships of Cl^- with **(a)** $\delta^{18}O$ and **(b)** δ^2H

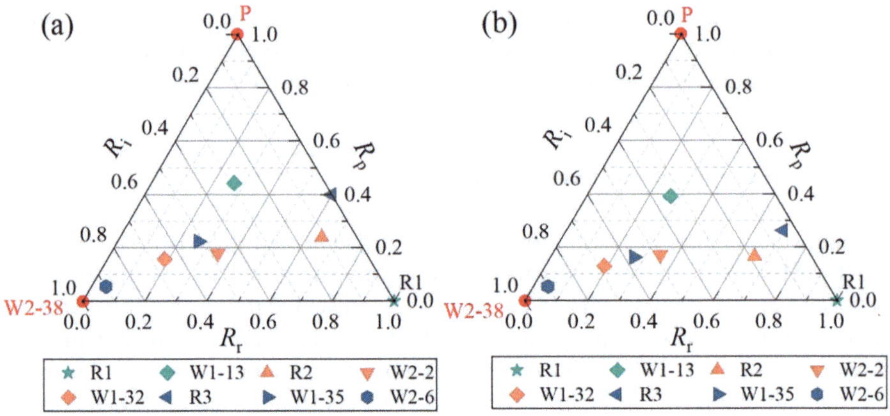

Fig. 7.8 Contribution ratios by the three end-member models (P in the upper angle is local precipitation, W2-38 in the lower left angle represents groundwater inflow, and the sample R2-1 located in the lower right denotes river water) established based on the relationships of Cl^- with (**a**) $\delta^{18}O$ and (**b**) δ^2H

basin, groundwater is commonly extracted for irrigation during April and May. A portion of this irrigation water percolates back into the subsurface, recharging the shallow aquifer and subsequently contributing to lateral groundwater inflow toward the river. Overall, lateral inflow is the dominant source of groundwater recharge along the Shidi River, particularly in the midstream and downstream reaches, followed by river seepage and precipitation.

7.4 Surface Water–Groundwater Exchange in the Xin River

Based on comprehensive geological data, including topographic and geomorphological information, and detailed field surveys conducted along the Xin River section, sufficient geological and hydrogeological information was obtained to meet the requirements for applying the CI method. Accordingly, the CI method was employed to investigate the surface water–groundwater connectivity of the Xin River.

The Xin River originates in the northern Qinling Mountains and flows northward through Huyi District, Xi'an City, eventually discharging into the Wei River at Majiazhai Village (Fig. 7.9). Due to intensive water resource development in the upstream region, the river segment below the mountain pass has become seasonal, typically lacking natural flow except during flood periods.

To assess surface water–groundwater interactions, 25 sampling sites were selected, evenly distributed across the upstream, midstream, and downstream sections. Groundwater samples were collected from shallow wells (<30 m deep) located within 1 km of the river channel. The dataset includes ten surface water

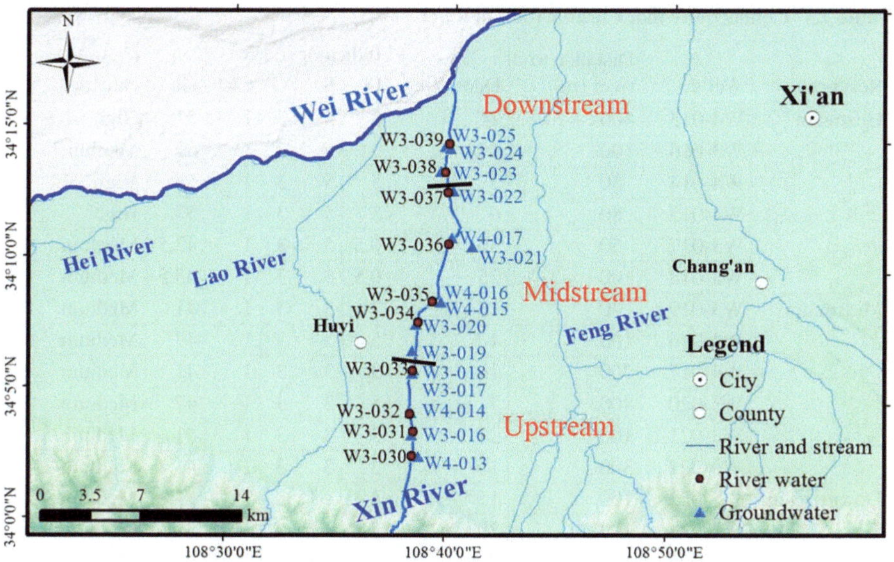

Fig. 7.9 Surface water and groundwater sampling points in Xin River basin

samples (Upstream: R3-030, R3-031, R3-032, R3-033; Midstream: R3-034, R3-035, R3-036, R3-037; Downstream: R3-038, R3-039) and 15 groundwater samples (Upstream: W3-016, W3-017, W3-018, W4-013, W4-014, W4-015; Midstream: W3-019, W3-020, W3-021, W3-022, W4-016, W4-017; Downstream: W3-023, W3-024, W3-025).

The indicators and results used to evaluate surface water–groundwater connectivity along the Xin River section, based on the CI method, are summarized in Table 7.5. At the mountain pass outlet in the upstream section, the riverbed sediments consist predominantly of gravel and coarse sand, resulting in a high permeability score of 5 (Fig. 7.10). In the midstream section, the sediments transition to silt and silty sand, yielding a moderate score of 3. Further downstream, the riverbed is composed of finer materials such as silt and clay loam, which correspond to a low permeability score of -1.

All shallow wells surveyed were located within 1000 m of the river, with distances ranging from 50 to 700 m; notably, 99.6% of the wells were within 500 m. In the upstream section, groundwater in the shallow aquifer typically occurs at greater depths, resulting in lower scores (3 or 0.5). However, localized geological anomalies can lead to significantly shallower water tables. For instance, samples W4-013 and W4-014 are situated near a hot spring area, where groundwater is found at depths of only 2 m, yielding high connectivity scores of 5.

In the midstream section, samples W3-020 and W4-016 exhibit anomalously high water levels. Particularly, W4-016 is located just 100 m from the river, suggesting a zone of enhanced surface water–groundwater interaction. Downstream, the terrain becomes increasingly flat, which corresponds to generally shallower

Table 7.5 Connectivity index results for Xin River

Reaches	Wells	Distance to river (m)	Depth (m)	Indicator scores				CI	Connectivity category
				D	S	G	GM		
Upstream	W4-013	400	2	5	5	3	1	57	High
	W3-016	100	15	3	5	3	1	51	Medium
	W4-014	50	2	5	5	3	1	57	High
	W4-015	80	6	5	5	3	1	57	High
	W3-017	50	30	0.5	5	3	1	43.5	Medium
	W3-018	100	25	0.5	5	3	1	43.5	Medium
Midstream	W3-019	70	15	3	3	3	1	41	Medium
	W4-016	100	4.5	5	3	3	1	47	Medium
	W4-017	700	10	3	3	3	1	41	Medium
	W3-020	400	7	5	3	3	1	47	Medium
	W3-021	100	18	3	3	3	1	41	Medium
	W3-022	300	15	3	3	3	1	41	Medium
Downstream	W3-023	100	15	3	−1	3	1	21	Medium
	W3-024	100	20	3	−1	3	1	21	Medium
	W3-025	150	20	3	−1	3	1	21	Medium

Fig. 7.10 Xin River upstream, midstream, and downstream

groundwater depths and higher scores (3 or 5). The downstream geological formations are composed mainly of sand and clay, contributing to a moderate permeability score of 3.

The CI calculation results, presented in Table 7.5, indicate that the upstream segment of the Xin River exhibits the highest connectivity values, ranging from 43.5 to 57. This high connectivity is primarily attributed to the presence of coarse sand and gravel sediments, which enhance infiltration. Three upstream locations recorded CI values of 57, indicating strong surface water–groundwater connectivity, while three other sites showed medium connectivity (CI < 53), mainly due to deeper groundwater levels that reduce interaction potential.

In the midstream segment, the dominant sediment types are sandy loam, silt, and clay, particularly approaching the downstream boundary. These finer sediments result in moderately reduced connectivity, with CI values ranging from 41 to 47, classifying this segment as having medium connectivity. The downstream segment characterized by clay-rich sediments, which have low permeability and yield the lowest CI value of 21, although this still falls within the medium connectivity category.

Overall, sediment grain size decreases progressively from upstream to downstream, resulting in reduced infiltration capacity and diminishing surface water–groundwater interaction. Based on the CI method, the upstream segment of the Xin River demonstrates high connectivity with groundwater, while the midstream and downstream segments show medium connectivity.

7.5 Surface Water–Groundwater Exchange in the Qingshui River

Given the relatively limited geological and hydrogeological data available for the Qingshui River, the hydrochemical approach was selected for a qualitative assessment of surface water–groundwater connectivity (Fig. 7.11). This analysis was based on field sample data, with differences in chloride (Cl^-) concentrations used for a quantitative evaluation of hydraulic connection. For this purpose, five water samples were collected from the Qingshui River basin, including three surface water samples (mid-upstream: W4-028; mid-downstream: W4-032 and W4-030) and two groundwater samples (mid-upstream: W4-027; mid-downstream: W4-029). The locations of the sampling sites are shown in Fig. 7.11.

Durov diagram analysis (Fig. 7.12) reveals that both surface water and groundwater in the Qingshui River study area share a common hydrochemical type, i.e., HCO_3–Ca·Mg type. Samples collected from the mid-upstream and mid-downstream sections cluster closely on the diagram, indicating a likely hydraulic connection between the river and adjacent groundwater. Notably, surface water and

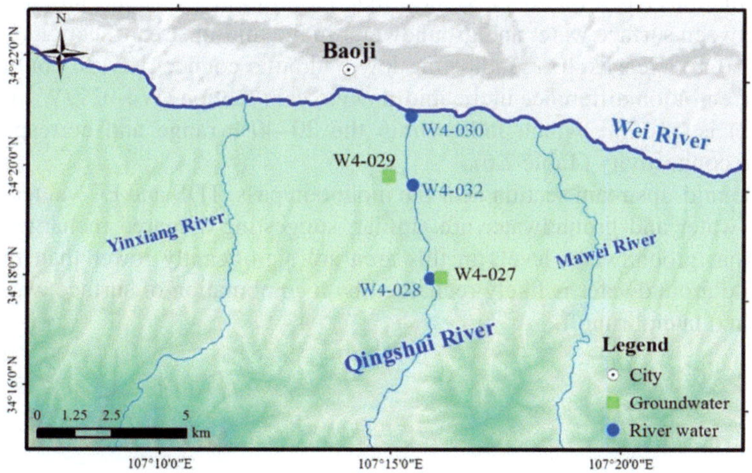

Fig. 7.11 Sampling points in the Qingshui River

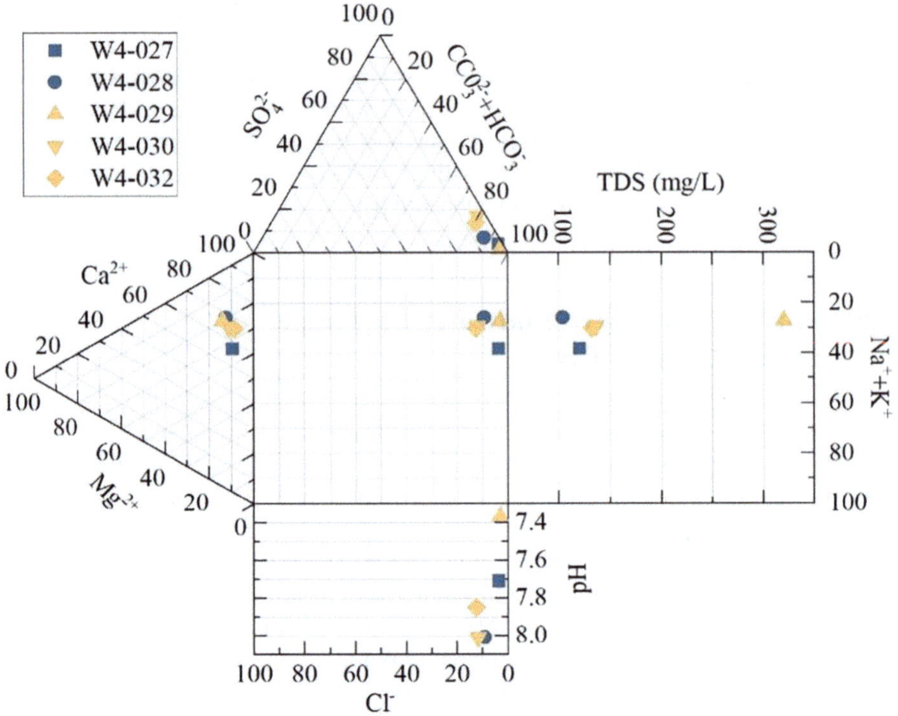

Fig. 7.12 Hydrochemical diagram of surface water and groundwater

groundwater samples from the mid-downstream section exhibit nearly identical dominant ion compositions, suggesting a stronger degree of connectivity in this part of the river system.

Calculations based on Eq. (7.3) show that the chloride (Cl^-) concentration difference between surface water and groundwater in the mid-upstream section (W4-027 and W4-028) exceeds 40%, indicating low hydraulic connectivity. In contrast, the Cl^- concentration difference in the mid-downstream section (W4-029, W4-030 and W4-032) is 28.57%, which falls within the 20–40% range and corresponds to medium connectivity (Table 7.6).

In the mid-upstream section near the mountain pass, TDS and EC values in both surface water and groundwater are similar, suggesting a shared recharge source. Given that groundwater levels in this area are significantly lower than the river stage, the groundwater is likely recharged by a combination of surface water infiltration and lateral runoff.

Table 7.6 Connectivity index determined by Cl^- concentration

Sections	Mean Cl^- of groundwater (mg/L)	Mean Cl^- of river water (mg/L)	Concentration difference (%)	Connectivity
Mid-upstream	2	4	66.67	Medium
Mid-downstream	8	6	28.57	Medium

In the mid-downstream section, TDS, EC, and Cl^- concentrations in both surface water and groundwater increase notably compared to mid-upstream levels. This trend is likely driven by strong evaporative processes that concentrate solutes in surface water as it flows downstream. However, the concentrations of TDS, EC, and Cl^- in groundwater are substantially higher than those in surface water in this section. This disparity indicates that, along the flow path from upstream to downstream, groundwater is subject to intense water–rock interaction processes. The geochemical effects of these interactions exceed those caused by evaporation alone, leading to greater solute enrichment in the groundwater.

7.6 Summary

This chapter investigates the exchange between surface water and groundwater in the study areas using CI, hydrochemical and isotopic methods, and cluster analysis. A comprehensive assessment in the Shidi River basin confirmed that these methods provide consistent results for evaluating hydraulic connectivity. Consequently, only the most appropriate methods were applied to each segment of the Xin and Qingshui Rivers.

In the Shidi River basin, analyses indicate a medium hydraulic connection between rivers and groundwater. EMMA reveals that lateral groundwater inflow is the dominant recharge source, especially midstream and downstream, followed by river seepage and precipitation.

For the Xin River basin, the CI method shows high surface water–groundwater connectivity upstream, while the midstream connectivity is medium. From midstream to downstream, medium connectivity is prevalent due to low permeability of river sediment.

In the Qingshui River basin, Durov diagrams and chloride concentration differences indicate medium connectivity mid-downstream and low connectivity mid-upstream. In the mid-downstream section, solute concentrations in both surface water and groundwater increase due to strong evaporation. However, groundwater shows significantly higher concentrations, reflecting intense water–rock interactions that surpass evaporative effects in enriching solutes.

References

Bu JW, Liu W, Zhao P, Ling K (2020) Comparative study of hydrochemical classification based on different hierarchical cluster analysis methods. Int J Environ Res Public Health 17(24):9515. https://doi.org/10.3390/ijerph17249515

Christophersen N, Neal C, Hooper RP et al (1990) Modelling stream water chemistry as a mixture of soil water endmembers-a step towards second-generation acidification models. J Hydrol 116(1):307–320. https://doi.org/10.1016/0022-1694(90)90130-P

Hooper RP, Christophersen N, Peters NE (1990) Modelling stream water chemistry as a mixture of soil water end-members-an application to the Panola Mountain catchment, Georgia USA. J Hydrol 116(1):321–343. https://doi.org/10.1016/0022-1694(90)90131-G

Li PY, Wu JH, Qian H (2016) Preliminary assessment of hydraulic connectivity between river water and shallow groundwater and estimation of their transfer rate during dry season in the Shidi River, China. Environ Earth Sci 75:99. https://doi.org/10.1007/s12665-015-4949-7

Li P, Wu J, Tian R, He S, He X, Xue C, Zhang K (2018) Geochemistry, hydraulic connectivity and quality appraisal of multilayered groundwater in the Hongdunzi Coal Mine, Northwest China. Mine Water Environ 37(2):222–237. https://doi.org/10.1007/s10230-017-0507-8

Li XF, Li PY, Liu JF (2025) Water security in the Qinling Mountains: current status, climate and human influences, and adaptive strategies. Int J Water Resour Dev. https://doi.org/10.1080/07900627.2025.2478387

Liu Y, Yamanaka T (2012) Tracing groundwater recharge sources in a mountain–plain transitional area using stable isotopes and hydrochemistry. J Hydrol 464–465:116–126. https://doi.org/10.1016/j.jhydrol.2012.06.053

Lu H, Li PY, Wang D (2025) Hydrochemical characteristics and transformation relationship of surface water and groundwater in the Hua County, Guanzhong Plain (China). Water Environ Res 97(6):e70115. https://doi.org/10.1002/wer.70115

Mohammed N, Celle JH, Huneau F et al (2014) Isotopic and geochemical identification of main groundwater supply sources to an alluvial aquifer, the Allier River valley (France). J Hydrol 508:181–196. https://doi.org/10.1016/j.jhydrol.2013.10.051

Navarro M, Garcia AS, Sanchez M et al (2017) Radionuclides as natural tracers of the interaction between groundwater and surface water in the River Andarax, Spain. J Environ Radioact 180:9–18. https://doi.org/10.1016/j.jenvrad.2017.09.015

Oyarzu'n R, Barrera F, Salazar P et al (2014) Multi-method assessment of connectivity between surface water and shallow groundwater: the case of Limarí River basin, north-central Chile. Hydrogeol J 22:1857–1873. https://doi.org/10.1007/s10040-014-1170-9

Ren XF, Li PY, He XD et al (2024) Tracing the sources and evaporation fate of surface water and groundwater using stable isotopes of hydrogen and oxygen. Sci Total Environ 931:172708. https://doi.org/10.1016/j.scitotenv.2024.172708

Shu L, Gong R, Luan J et al (2022) A integrated method to quantify flow exchanges between surface water and groundwater: take Suzhou section of the Xinbian River as an example. Adv Water Sci 33(1):57–67. (In Chinese with English Abstract)

Tao J (2012) Coupled modeling of surface water and groundwater in Shayinghe River Basin. Master's thesis, Hefei University of Technology (In Chinese)

Terzer-Wassmuth S, Araguás-Araguás LJ, Wassenaar LI et al (2023) Global and local meteoric water lines for $\delta^{17}O/\delta^{18}O$ and the spatiotemporal distribution of $\Delta'^{17}O$ in Earth's precipitation. Sci Rep 13:19056. https://doi.org/10.1038/s41598-023-45920-8

Wang Y, Yin D, Wang X et al (2021) Groundwater–surface water interactions in the Baiyangdian wetland, Xiong'an New Area and its impact on reed land. Geol China 48(5):1368–1381. (In Chinese with English Abstract)

West AG, February EC, Bowen GJ (2014) Spatial analysis of hydrogen and oxygen stable isotopes ("isoscapes") in ground water and tap water across South Africa. J Geochem Explor 145:213–222. https://doi.org/10.1016/j.gexplo.2014.06.009

Xu F, Li P, Wang Y, Du Q (2023) Integration of hydrochemistry and stable isotopes for assessing groundwater recharge and evaporation in pre- and post-rainy seasons in Hua County, China. Nat Resour Res 32(5):1959–1973. https://doi.org/10.1007/s11053-023-10235-y

Yang SH, Qian H, Xu PP et al (2025) Hydrogeochemical mechanism and interconversion processes of groundwater–surface water in the Chan River Basin: a new perspective from hydrochemistry and isotopes. J Environ Sci 157:890–907. https://doi.org/10.1016/j.jes.2025.02.003

Zhang L (2022) Hydraulic connection between river water and river bank groundwater in the Weinan section of the Weihe River, China. Master's thesis, Chang'an University (In Chinese)

Zhang Y, Su X, Wang Q et al (2020) Surface water–groundwater interactions in the western plain of the Ili Valley. J Beijing Normal Univers (Nat Sci) 56(5):664–674. (In Chinese with English Abstract)

Zhang L, Li P, He X (2022a) Interactions between surface water and groundwater in selected tributaries of the Wei River (China) revealed by hydrochemistry and stable isotopes. Hum Ecol Risk Assess 28(1):79–99. https://doi.org/10.1080/10807039.2021.2016054

Zhang Z, Duan K, Liu H et al (2022b) Runoff projections of the Qinling Mountains and their impact on water demand of Guanzhong region in Northwest China. J Mt Sci 19:2272–2285. https://doi.org/10.1007/s11629-022-7358-x

Zhao J, Liu GD, Meng YC et al (2021) Using stable isotopes as tracer to investigate hydrological condition and estimate water residence time in a plain region, Chengdu, China. Sci Rep 11:2812. https://doi.org/10.1038/s41598-021-82349-3

Chapter 8
Land Use Changes in Selected Areas of the Northern Qinling Mountains

Abstract The northern Qinling Mountains witnessed significant human activities, changing land use/land cover. The changes in land use can alter the processes of water circulation, thus affecting its quality. This chapter investigates land use and land cover (LULC) changes in selected areas of the northern Qinling Mountains from 2000 to 2020 and projects future trends up to 2040. Using remote sensing data and the GeoSOS-FLUS model, the study analyzes spatial distribution characteristics, structural changes, and transfer processes of six land use types: farmland, forest, grassland, urban and rural residential land, water, and wasteland. Results indicate that farmland remains the dominant land use type, though its area has decreased by 4% over two decades. Urban and rural residential land and water areas have expanded significantly, driven by economic development and environmental policies. The FLUS model predicts continued declines in farmland by 2040, with further growth in residential and water areas. These findings provide valuable insights for regional land use planning, ecological conservation, and sustainable development in the northern Qinling Mountains.

Keywords Land use/land cover · Human activities · Land use type · FLUS model · Land use change

8.1 Introduction

Land is the fundamental carrier for human ecological civilization, and it serves as an effective pathway for realizing the interaction between human and nature (Han et al. 2025). The structural layout of Land Use and Land Cover (LULC) change can directly affect regional ecosystems (Zhu et al. 2022). Over the long term, large-scale activities such as afforestation, conversion of farmland to forests, construction of water conservancy projects, and irrigation and drainage have accelerated the pace of industrialization and urbanization. This has also led to changes in land use types, structures, and coverage patterns, thereby impacting the surrounding ecological environment and biodiversity (Usman et al. 2015).

P. Li et al., *Water Resources Research in Northern Qinling Mountains, China*, Springer Hydrogeology, https://doi.org/10.1007/978-3-032-09395-0_8

In recent years, LULC in the northern Qinling Mountains has been constantly changing. For example, as the economic center of northwest China, Xi'an has witnessed rapid economic development. With the establishment of Xi'an-Xianyang New Area, significant changes have occurred in land use in the counties and districts surrounding Xi'an. The northern Qinling Mountains, known as the "National Central Park" and "China's Green Lung", functions as a barrier for ecological protection in the Guanzhong region (Li et al. 2022). Therefore, this book selected a part of the northern Qinling Mountains (Qishan County—Mei County—Zhouzhi County—Huyi District) as the specific study area (Fig. 8.1) to investigate the land use changes. In this selected study area, Qishan County only occupies a small portion at the westernmost edge.

Over the three decades, the LULC pattern in this region has undergone tremendous changes, with a large amount of farmland converted into kiwifruit orchards, making it the world's largest kiwifruit production base (Li et al. 2025). Existing studies have shown that the degree of soil erosion in orchard land is higher than in farmland (Li et al. 2021), which not only affects soil and rivers but also leads to the deterioration of the groundwater environment. Thus, it is particularly crucial to conduct in-depth research on the characteristics of land use changes in this area. This

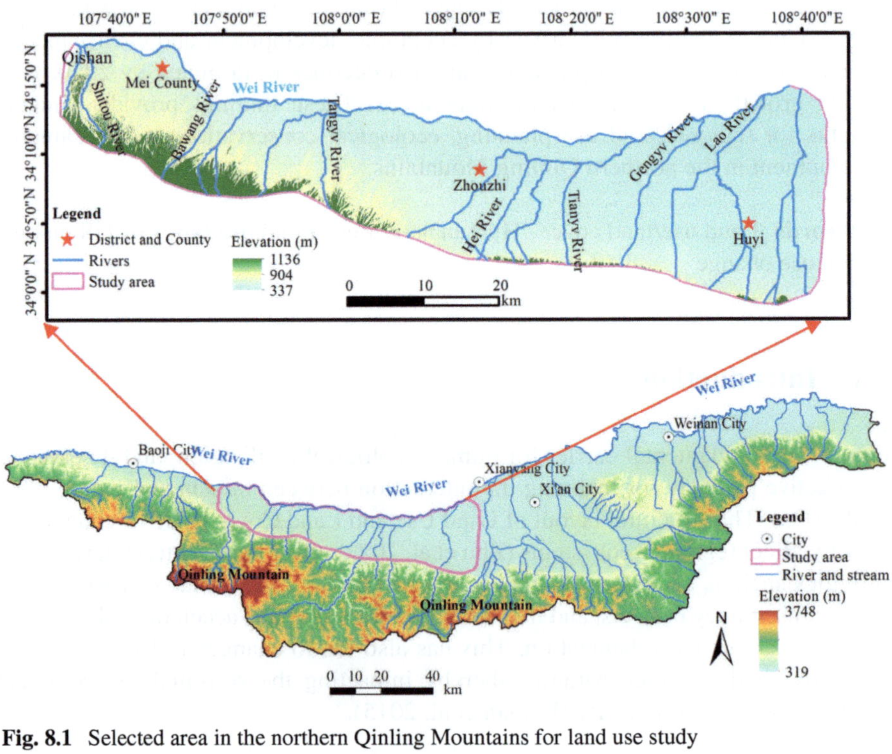

Fig. 8.1 Selected area in the northern Qinling Mountains for land use study

Table 8.1 Land use classifications for the northern Qinling Mountains

Land use types	Interpretation
Farmland	Land engaged in agricultural activities, which is mainly used for growing crops, with scattered fruit trees
Forest	Forestry land where arbors, shrubs, bamboos, etc., grow, including natural woodlands and artificial tree plantations
Grassland	Various types of grasslands dominated by herbaceous plants
Water	Natural terrestrial water areas and land for water conservancy facilities, including land occupied by rivers, reservoirs, lakes, and ponds
Urban and rural residential land	Land for urban and rural settlements, as well as land for industrial and mining operations, transportation, and other related purposes
Wasteland	Include bare land that has been leveled due to construction needs or other factors but remains unused, as well as exposed rock masses

can provide references for solving regional ecological problems, lay a theoretical foundation for the rational allocation of land resources and effective improvement of the ecological environment in the northern Qinling Mountains, assist government departments in formulating sound land use plans, and promote the positive development of the regional ecology.

This chapter is based on the remote sensing data of five periods (2000, 2005, 2010, 2015, and 2020). The characteristics of land use changes in selected areas of the northern Qinling Mountains over the past 20 years were analyzed. Meanwhile, based on the GeoSOS-FLUS model, this chapter simulates future land use, which can help understand land use changes under the influence of human activities and natural factors as well as future land use scenarios (Li et al. 2017). Using the remote sensing data of land use types in 2000 and 2020 as the foundation, the simulation data that comprehensively considers economic and environmental development is selected for analysis to simulate the land use distribution in 2040.

According to the land use categories specified in *Classification of Current Land Use* (GB/T 21010-2017), the land use types for the study area can be classified into six types: farmland, forest, grassland, urban and rural residential land, water, and wasteland. The definitions of each type are presented in Table 8.1.

8.2 Spatial Distribution of Land Use

8.2.1 Data Sources

The land use data for the five periods of 2000, 2005, 2010, 2015, and 2020 used in the study area were obtained from the Resource and Environment Science Data Center of the Chinese Academy of Sciences (http://www.resdc.cn), with a spatial resolution of 30 m × 30 m and a consistent spatial coordinate system of "GCS_Xian_1980".

8.2.2 Analysis of Spatial Distribution Characteristics

The land use of the selected study area from 2000 to 2020 is illustrated in Fig. 8.2. The study area displayed analogous land use patterns during this period. Farmland remained the dominant type, with its area proportion gradually decreasing from 76.3% in 2000 to 72.1% in 2020. Grassland and forest were sporadically distributed in Mei County and the southern part of the study area. From 2000 to 2020, the proportion of forest area varied between 3.4 and 3.5%. Due to the ecological protection policies for the Qinling Mountains, forests were less affected by human disturbances. Similarly, the proportion of grassland area fluctuated between 5.5 and 6.1%. Urban and rural residential land was scattered within farmlands, with its proportion gradually increasing from 11.4 to 14.7% during 2000–2020, reflecting the conversion of some farmland to residential land in the process of urbanization and rural construction. Huyi District, being close to the main urban area of Xi'an, experienced accelerated surrounding development driven by urbanization, resulting in a significant expansion of residential areas. The proportion of water area was relatively small in 2000, accounting for 2.7%. After 2005, the area increased and stabilized, ranging between 3.4% and 3.7%. Spatially, the water area along the Wei River increased, which is not only attributed to the ecological governance projects of the Wei River but also related to the regional strict water resource management system and the guarantee of ecological base flow in the river course. Meanwhile, this is also consistent with Xi'an's practice of promoting the construction of the "Three Rivers and One Mountain" ecological corridor in 2017 (Yang et al. 2024). In addition, the

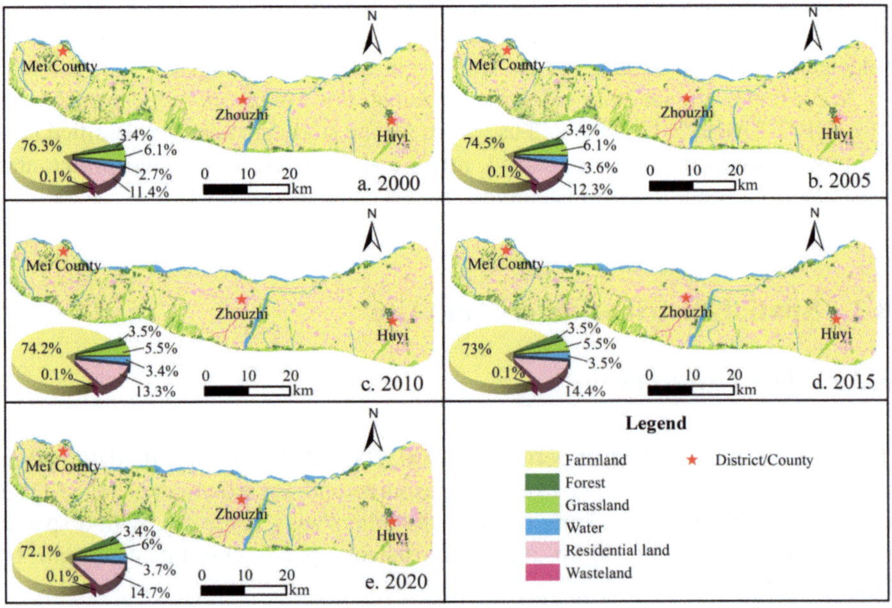

Fig. 8.2 Land use patterns and area proportions in the study area

proportion of limited wasteland area remained at 0.1%, mainly distributed in the central part of Zhouzhi County.

8.3 Characteristics of Land Use Type Changes

8.3.1 Study Method

8.3.1.1 Magnitude and Dynamic Degree of Land Use Change

The magnitude of land use type change (R) can quantitatively reflect the change of a specific land use type over a certain period, and its calculation formula is as follows (Eq. 8.1):

$$R = \frac{S_2 - S_1}{S_1} \times 100\% \tag{8.1}$$

Where, S_1 and S_2 represent the area of a specific land use type at the beginning and end of the study period, respectively.

The dynamic degree (K) of a given land use type indicates the rate of change of a certain land use type, which is calculated by Eq. (8.2):

$$K = \frac{S_2 - S_1}{S_1} \times \frac{1}{T} \times 100\% \tag{8.2}$$

Where, T denotes the research time, usually expressed in year. S_1 and S_2 are the same as above.

8.3.1.2 Land Use Transfer Matrix

The land use transfer matrix is a commonly used method for analyzing the quantity and direction of changes in land use types. It can trace the destinations and sources of transfer for various land use types. The expression of the land use transfer matrix is as follows (Eq. 8.3):

$$M = \begin{bmatrix} M_{11} & M_{12} & \cdots & M_{1j} & \cdots & M_{1n} \\ M_{21} & M_{22} & \cdots & M_{2j} & \cdots & M_{2n} \\ \vdots & \vdots & \ddots & \vdots & \vdots & \vdots \\ M_{i1} & M_{i2} & \cdots & M_{ij} & \cdots & M_{in} \\ \vdots & \vdots & \ddots & \vdots & \vdots & \vdots \\ M_{n1} & M_{n2} & \cdots & M_{nj} & \cdots & M_{nn} \end{bmatrix} \tag{8.3}$$

Where, n represents the number of land use types, while i and j denote the land use types at the beginning and end of the study, respectively. Each column element in the matrix indicates the area of a specific land use type converted from other types. Each row element represents the area converted out from a specific land use type to other types. The diagonal elements in this matrix represent the area of land use types that remain unchanged without transfer.

8.3.2 *Land Use Changes in the Selected Area of the Northern Qingling Mountains*

The area and proportions of various land use types in the selected area of the northern Qinling Mountains in 2000, 2005, 2010, 2015, and 2020 were statistically analyzed, with the magnitude of change and dynamic degree during 2000–2020 computed as well, as presented in Table 8.2 and Fig. 8.3.

Although the land use from 2000 to 2020 illustrated in Fig. 8.3 and Table 8.2 exhibits similar spatial distribution characteristics, variations exist in the area and proportion of land use across different years. A comparison of the area of each land use type between 2000 and 2020 reveals that farmland is the primary land use type in the northern Qinling Mountains. Its area in 2000 and 2020 was 1347.07 km^2 and 1273.80 km^2, respectively, accounting for 76% and 72% of the total area of the selected study region. The area of farmland decreased by 73.27 km^2 over the 20 years. The area of forest and grassland changed slightly, from 60.32 km^2 and 107.81 km^2 in 2000 to 60.47 km^2 and 106.26 km^2 in 2020, respectively. Water increased from 47.11 km^2 in 2000 to 64.65 km^2 in 2020, with a growth of 17.54 km^2. Urban and rural residential land expanded from 201.53 km^2 in 2000 to 258.66 km^2 in 2020, with an increase of 57.13 km^2.

Table 8.2 Changes in land use structures in the northern foot of the Qinling Mountains from 2000 to 2020

Years and statistics		Farmland	Forest	Grassland	Water	Residential land	Wasteland
2000	Area (km^2)	1347.07	60.32	107.81	47.11	201.53	2.23
	Proportion (%)	0.76	0.03	0.06	0.03	0.11	0
2005	Area (km^2)	1316.27	60.23	107.16	63.59	216.58	2.23
	Proportion (%)	0.75	0.03	0.06	0.04	0.12	0
2010	Area (km^2)	1309.35	62.01	97.61	59.7	235.17	2.23
	Proportion (%)	0.74	0.4	0.06	0.03	0.13	0
2015	Area (km^2)	1288.78	61.31	96.61	61.52	255.62	2.23
	Proportion (%)	0.73	0.03	0.05	0.03	0.14	0
2020	Area (km^2)	1273.8	60.47	106.26	64.65	258.66	2.23
	Proportion (%)	0.72	0.03	0.06	0.04	0.15	0
2000–2020	R (%)	−5.44	0.25	−1.44	37.23	28.35	0
	K (%)	−0.27	0.01	−0.07	1.86	1.42	0

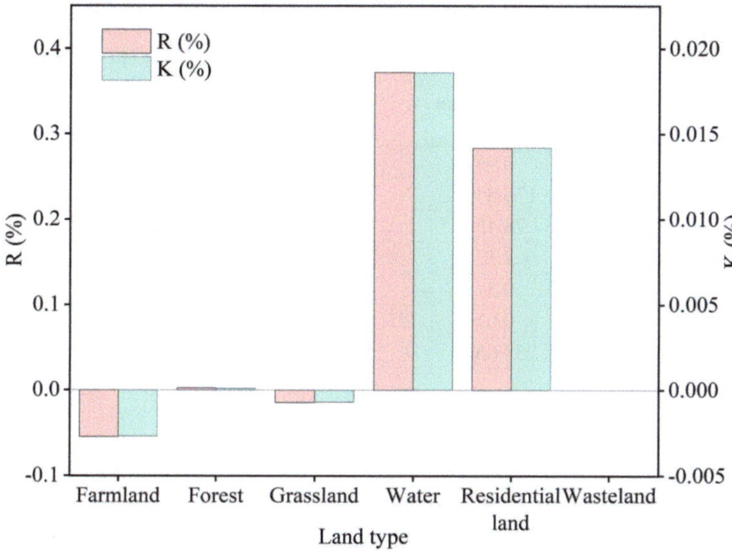

Fig. 8.3 Magnitude of change (*R*) and dynamic degree (*K*) of land use in the northern Qinling Mountains from 2000 to 2020

The magnitude of change and dynamic degree of land use change can reveal the relative intensity of changes for various land use types, which are significantly influenced by the initial area of the land type. In Table 8.2 and Fig. 8.3, the absolute values of the magnitude of change and dynamic degree (annual change rate) of land use in the northern Qinling Mountains from 2000 to 2020 are as follows: water > urban and rural residential land > farmland > grassland > forest > wasteland. Among them, water exhibits the most significant changes, with the magnitude of change and dynamic degree reaching 37.23% and 1.86%, respectively, followed by urban and rural residential land with the magnitude of change and dynamic degree being 28.35% and 1.42%, respectively. The total area of wasteland has remained unchanged over the years, thus its magnitude of change and dynamic degree are 0.

8.3.3 *Land Use Transfer and Spatial Change*

To understand the specific area of each land use change within the region from 2000 to 2020, intersect operation was performed on the vector data of land use classification in the northern Qinling Mountains for 2000 and 2020 using ArcGIS 10.3. Based on the operation results, the land use transfer matrix (Table 8.3), land use transfer distribution map (Fig. 8.4), and land use transfer chord diagram (Fig. 8.5) of the northern Qinling Mountains from 2000 to 2020 were obtained. Among them, farmland, water, and urban and rural residential land showed relatively significant changes. Specifically, from 2000 to 2020, 8.05 km^2 of farmland was converted to

Table 8.3 Land use transfer matrix of the selected area in the northern Qinling Mountains from 2000 to 2020

	2020						
2000	Grassland	Residential land	Farmland	Forest	Water	Wasteland	Total
Grassland	95.36	0.93	6.72	1.46	3.28	/	107.76
Residential land	0.54	176.85	23.52	0.36	0.21	0.03	201.52
Farmland	8.02	78.45	1235.15	6.72	18.27	0.37	1346.98
Forest	1.31	2.12	5.03	51.41	0.41	/	60.29
Water	0.98	0.3	2.92	0.5	42.34	/	47.03
Wasteland	0.01	0.01	0.39	/	/	1.83	2.23
Total	106.22	258.65	1273.75	60.44	64.52	2.23	1765.81

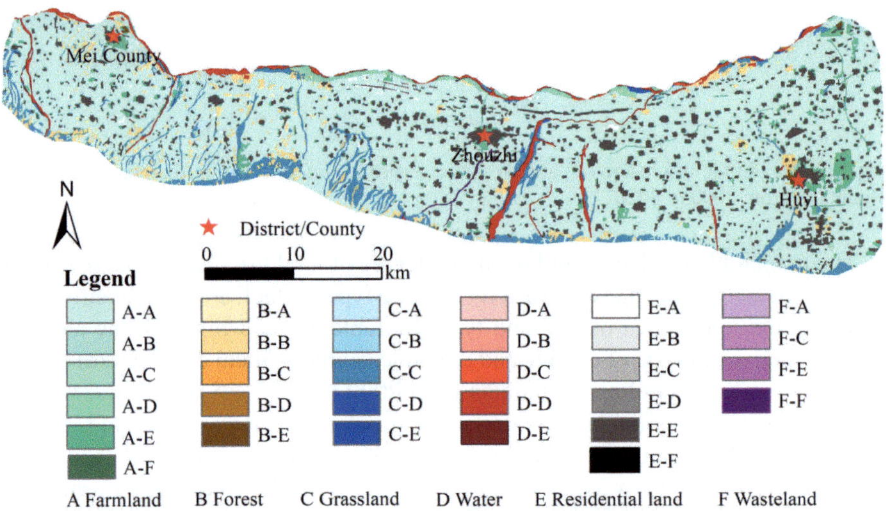

Fig. 8.4 Distribution map of land use transfer in the study area from 2000 to 2020

Fig. 8.5 Chord plot of land use transfer from 2000 to 2020

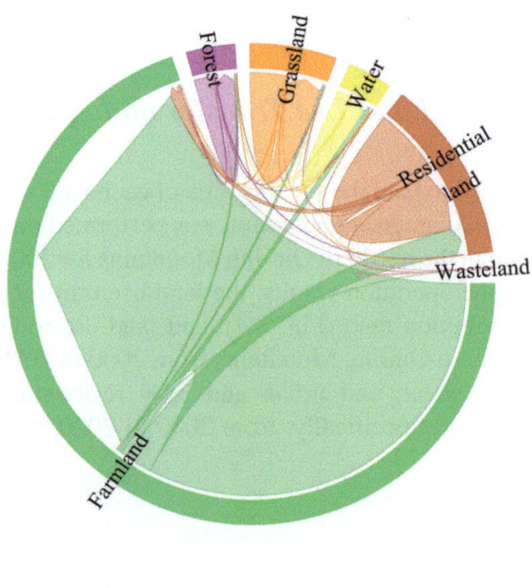

grassland, 78.46 km^2 to residential land, 6.73 km^2 to forest, and 18.31 km^2 to water. 6.77 km^2 of grassland was converted to farmland, 0.93 km^2 to residential land, 1.46 km^2 to forest, and 3.28 km^2 to water; 0.54 km^2 of residential land was converted to grassland, 23.52 km^2 to farmland, 0.36 km^2 to forest, and 0.22 km^2 to water. 1.31 km^2 of forest was converted to grassland, 2.12 km^2 to residential land, 5.04 km^2 to farmland, and 0.41 km^2 to water; 0.98 km^2 of water was converted to grassland, 0.30 km^2 to residential land, 0.50 km^2 to forest, and 2.92 km^2 to farmland. Figure 8.5 showed that from 2000 to 2020, the area of farmland in the study area generally showed a decreasing trend, while the areas of water and residential land showed an increasing trend over these 20 years.

8.4 Future Changes in Land Use Types

8.4.1 Data Sources

In this section, the spatial distribution dataset of Chinese gross domestic product (GDP) in 2020, spatial population distribution data, the digital elevation model (DEM) of Shaanxi Province with a 30 m resolution, land use data for two periods (2000 and 2020), average temperature dataset with 1 km resolution and average precipitation dataset from the Resource and Environment Science Data Center of China were used to predict the future changes in land use. The Chinese 2020 highway and railway datasets from the National Geomatics Center of China was also adopted. All raster data used in this section were resampled to a unified resolution of 30 m × 30 m.

8.4.2 Construction of the FLUS Model

The FLUS model is a combined model for simulating future land use changes, consisting of two major modules: the Artificial Neural Network (ANN)-driven probability module and the Cellular Automaton (CA) module (Yang et al. 2023). Firstly, the model employs ANN technology to calculate the occurrence probability of each pixel for each land use type. Secondly, the CA module with an adaptive inertial competition mechanism is used to reflect the competition among different land use types. It mainly calculates the conversion and distribution of each land use type in each sub-unit based on occurrence probability, the influence of inter-cell neighborhoods, conversion costs, and inertial coefficients, thereby achieving higher-precision simulation of land use changes. Finally, the Kappa coefficient is utilized to measure the applicability of the model.

8.4.2.1 Principle of FLUS Model Construction

1. Probability of driving based on ANN

   ANN is an information processing system that is based on simulating the layout of the human brain's neural network, while generalizing, abstracting, and simulating it (Zhao et al. 2022). A neural network is a computing system composed of numerous interconnected processing units, i.e., neurons, which constructs a three-layer organizational framework (input layer, hidden layer, and output layer) in the form of a network. Its functional characteristics are influenced by the network layout, connection density, and neuron processing methods. The model uses land cover data of the baseline period and natural/socio-economic factors as basic input data to output the probability of suitability in land use types. The input neurons are expressed as Eq. (8.4):

$$X = \left(x_1, x_2, \ldots, x_n\right) \tag{8.4}$$

Where, x_i denotes the i-th neuron in the input layer, with a total of n neurons, representing the influence of driving factors.

The interactive information between neurons in the input layer and the hidden layer is shown in Eq. (8.5):

$$N_j\left(p,q\right) = \sum_i w_{ij} \times x_i\left(p,q\right) \tag{8.5}$$

Where, N_j (p, q) represents the information perceived by neuron j in the hidden layer. x_i (p, q) denotes the information transmitted by the i-th neuron in the input layer after the q-th training of cell p. w_{ij} is the weight value between neurons i and j during neural learning.

The activation relationship between the hidden layer and the output layer is expressed as Eq. (8.6):

$$S\left(N_j\left(p,q\right)\right) = \frac{1}{1 + e^{-N_j(p,q)}} \tag{8.6}$$

Where, S represents the Sigmoid function.

The probability of suitability for the k-th land use type after cell p experiencing q-th iteration is expressed as Eq. (8.7):

$$P_{p,k}^{(q)} = \sum_j w_{jk} \times S\left(N_j\left(p,q\right)\right) = \sum_j w_{jk} \times \frac{1}{1 + e^{-N_j(p,q)}} \tag{8.7}$$

Where, w_{jk} denotes the weight value of neurons between the hidden layer and the output layer.

2. CA Simulation based on Adaptive Inertial Mechanism

   In the FLUS model, to determine the conversion and distribution of various land use types in the cell, the first step is to determine the domain effect, which can be expressed as Eq. (8.8):

$$E_{p,k}^{(q)} = \frac{\sum_{N\times N} con\left(c_p^{(q-1)} = k\right)}{N \times N - 1} \times w_k \tag{8.8}$$

Where, $\sum_{N\times N} con\left(c_p^{(q-1)} = k\right)$ represents the total number of pixels of land use type k after the completion of the $(q - 1)$-th iteration. N denotes the boundary length of the calculation interface. w_k is the degree of neighborhood influence among various land use types.

Secondly, the inertial coefficient can be expressed as Eq. (8.9):

$$I_k^{(q)} = \begin{cases} I_k^{(q-1)} & \left|D_k^{(q-2)}\right| \le \left|D_k^{(q-1)}\right| \\ I_k^{(q-1)} \times \dfrac{D_k^{(q-2)}}{D_k^{(q-1)}} & 0 > D_k^{(q-2)} > D_k^{(q-1)} \\ I_k^{(q-1)} \times \dfrac{D_k^{(q-1)}}{D_k^{(q-2)}} & D_k^{(q-1)} > D_k^{(q-2)} > 0 \end{cases} \tag{8.9}$$

Where, $I_k^{(q-1)}$ is the inertial coefficient of land use type k at the q-th iteration. $D_k^{(q-1)}$ represents the difference between the land use type k and the set value during the $(q - 1)$-th neural learning. If the difference is 0, the iteration stops. If the difference is not 0, the inertial coefficient is adjusted adaptively and dynamically until the difference becomes 0.

Finally, the conversion cost represents the difficulty of converting the current land use type to other land use types. If a certain type of land can be converted to other land use types, it is set to 1; otherwise, it is set to 0.

By integrating the above four factors, the expression for the total probability affecting the conversion of land use types is expressed as Eq. (8.10):

$$TP_{p,k}^{q} = P_{p,k} \times E_{p,k}^{q} \times I_k^{q} \times \left(1 - sc_{c\to k}\right) \tag{8.10}$$

Where, $TP_{p,k}^{q}$ denotes the probability that grid cell p converts from the initial land use type to the target land use type k after q iterations. $1 - sc_{c\to k}$ represents the conversion cost.

8.4.2.2 Steps for Constructing the FLUS Model

1. Selection of driving factors

   Land use changes are influenced by natural, social, economic, and other driving factors across different temporal and spatial scales, characterized by long-term dynamics, complex action modes, and highly intricate nonlinear features (Lai et al. 2025). Therefore, the accuracy of the model is closely linked to the selection of driving factors. Affected by both natural and socioeconomic factors, this section aims to objectively reflect the conversion processes between various land types. Considering the accessibility, consistency (consistency in time, space, coordinates, and row-column numbers), quantifiability, spatial variability, and correlation significance of factor data, 8 driving factors of land use change in the northern Qinling Mountains were selected. Specifically, natural factors include elevation, annual average temperature, annual average precipitation, and slope, while socioeconomic factors include population density, GDP, distance to railways, and distance to highways. Detailed information on each driving factor and their data sources are provided in Table 8.4 and Fig. 8.6.

   Based on the DEM data, raster maps of elevation (Fig. 8.6a) and slope (Fig. 8.6b) were generated, respectively, and then the Euclidean distances from each pixel to highways and railways within the selected region were calculated based on the distribution of highways and railways, resulting in the corresponding raster maps (Fig. 8.6c, d). Finally, the raster data for the selected region were extracted using ArcGIS, including annual average precipitation (Fig. 8.6e), annual average temperature (Fig. 8.6f), population density (Fig. 8.6g), and GDP (Fig. 8.6h).
2. Parameter setting

   (a) Driving probability of ANN

   First, the land use data of the baseline year 2000 was inputted. Second, the neural learning sampling parameter was set to 20, with the unit being 1/1000 of the number of valid grids in the study area, and the hidden layer

Table 8.4 Driving factors for land use change in the northern foot of the Qinling Mountains

Driving factor	Name	Source
Natural factor	DEM	China Resource and Environment Science and Data Center
	Annual average temperature	
	Annual average precipitation	
	Slope	Calculated based on DEM
Socioeconomic factors	Population density	China Resource and Environment Science and Data Center
	GDP	
	The distance from the railway	National Geospatial Information Center
	The distance from the road	

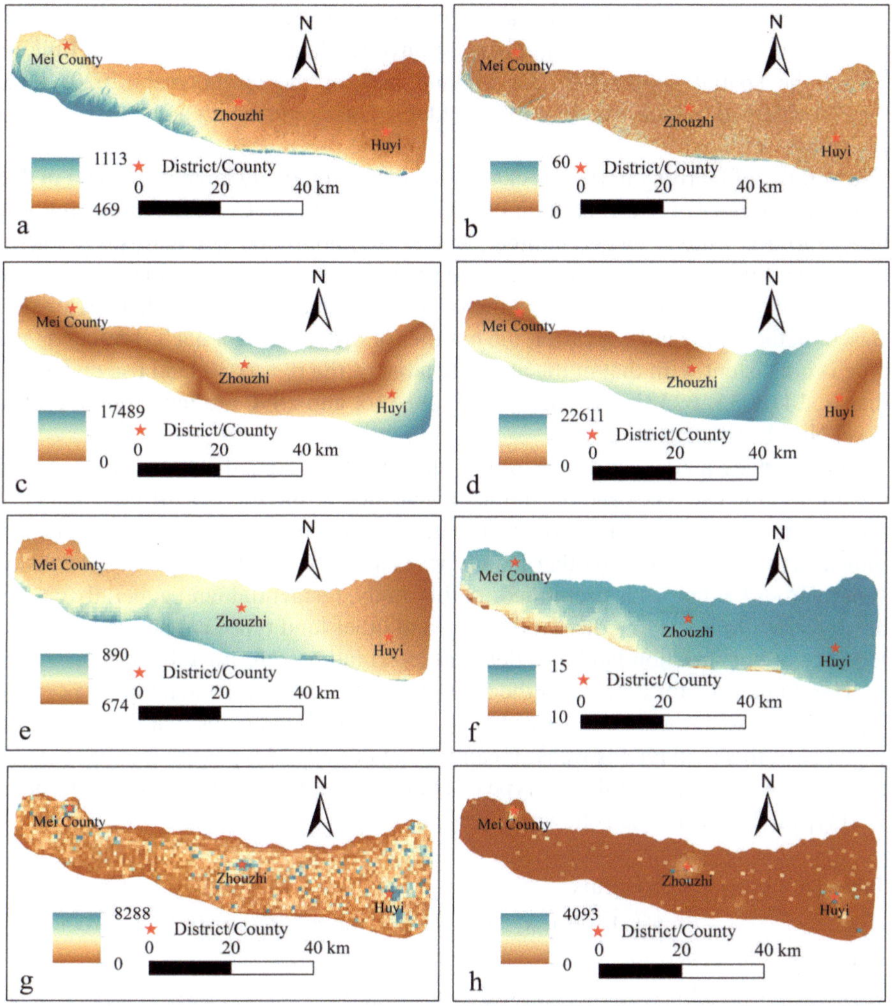

Fig. 8.6 Raster data of driving factors for land use change in the northern foot of the Qinling Mountains. (**a**) DEM; (**b**) Slope; (**c**) Distance to highways; (**d**) Distance to railways; (**e**) Precipitation; (**f**) Temperature; (**g**) Population density; (**h**) GDP

was set to 12 layers. Then, a total of 8 driving factors, including natural and socio-economic ones in the study area, were inputted and subjected to standardization processing. Finally, the suitability probability map was obtained by calculating the probability of each land use type appearing in each grid. The root mean square error was only 0.23, indicating a relatively small error in the calculation.

(b) CA module with adaptive inertial competition mechanism

The land classification raster data of the selected region in the northern Qinling Mountains in 2000 and its suitability probability map were used as the initial input data. Then, parameters such as the number of learning iterations, neighborhood value, acceleration factor, neighborhood weight, transition matrix, and target number of converted land were determined. Finally, the spatial simulation data of land use for the target year (2020) were obtained, which were used for model accuracy verification.

The number of learning iterations can be preset to a relatively large value, and in this study, the number of learning iterations is set to 300. In the Cellular Automaton, the neighborhood value is an odd number, and the neighborhood size in this study is set to 3. In addition, due to the relatively small size of the study area, the acceleration factor is set to 0.1. The first CA simulation aims to simulate the 2020 land use raster. Based on the 2000 land use raster, the number of raster cells for each simulated land type is consistent with that in the attribute table of the real 2020 land use raster file, that is, 1418035, 67323, 118162, 71405, 288866, and 2490 are input in the Future Pixel Number distribution.

Based on the actual conditions of the selected study area, the expansion capacity of various land use types and the mutual conversion costs between different land use types were set. In the conversion cost matrix, only 0 or 1 can be filled in, where 1 indicates that conversion to other land types is allowed, and 0 indicates that conversion to other land types is prohibited. In this study, residential land generally cannot be converted to water or other land types, and other land types cannot be converted to water or construction land. For detailed information, refer to Table 8.5.

The domain factor represents the strength of the expansion capacity of a certain type of land, with a value range of [0, 1]. A value of 0 indicates that the expansion capacity of this land type is weak, meaning that other land types are less likely to be converted to this type, while this type of land is more likely to be converted to other land types. The results are shown in Table 8.6.

Table 8.5 Conversion cost matrix

Type	Farmland	Forest	Grassland	Water	Residential land	Wasteland
Farmland	1	1	1	1	1	1
Forest	1	1	0	0	0	0
Grassland	1	1	1	1	1	1
Water	0	0	1	1	0	0
Residential land	0	0	1	0	1	0
Wasteland	1	1	1	0	0	1

Table 8.6 Parameter of domain factors

Type	Farmland	Forest	Grassland	Water	Residential land	Wasteland
Domain factor	1	0.7	0.5	0.4	0.6	0.1

8.4.3 *Prediction of Land Use Using the FLUS Model*

8.4.3.1 Model Performance

The land use status data in 2000 and adaptive probability map were imported into the cellular automaton module of the FLUS model. The model parameters were set as per previous introduction, and the target quantities of various land uses for 2020 were inputted. Finally, the cellular automaton module was run, and after a long-term iteration, the predicted land use in 2020 was obtained, as shown in Fig. 8.7.

Validation of the model performance was performed using the Kappa coefficient in the FLUS model, with 10% random sampling selected. The Kappa coefficient is an indicator for measuring classification accuracy calculated using a confusion matrix, which can effectively verify the simulation accuracy of the model. The closer the Kappa coefficient is to 1, the more accurate the results. Generally, a

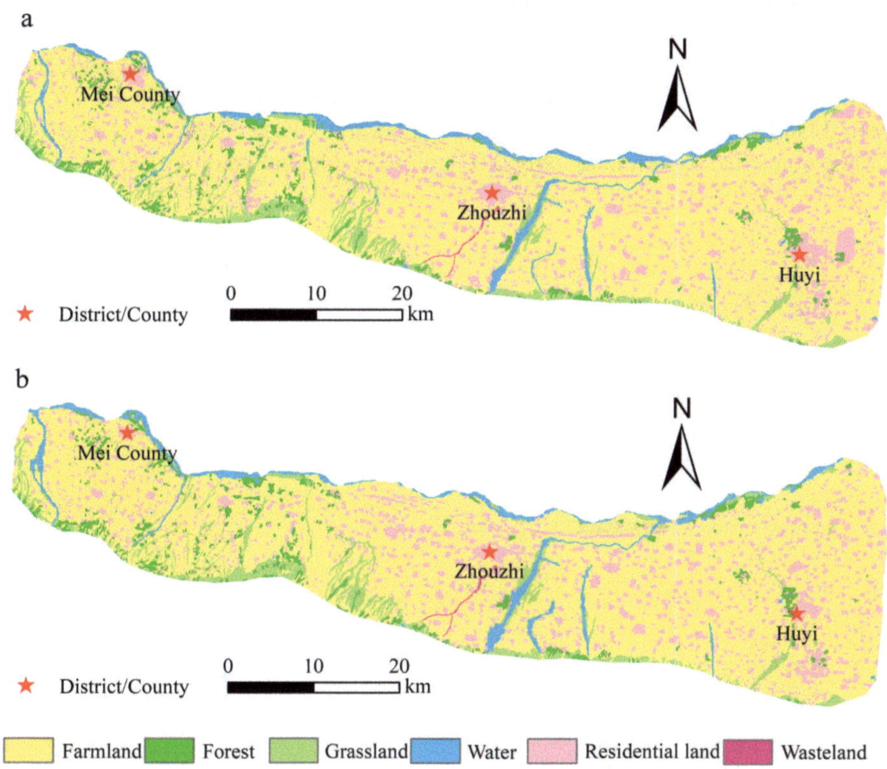

Fig. 8.7 Land use types. (**a**) Actual land use in 2020; (**b**) Predicted land use in 2020

Kappa coefficient greater than 0.75 indicates high simulation accuracy, while a Kappa coefficient between 0.5 and 0.75 indicates moderate model accuracy. In this simulation, the Kappa coefficient for the simulated and the real land use in 2020 was 0.769, indicating an overall good simulation performance of the model. Therefore, this model can be used for the prediction of land use in 2040.

8.4.3.2 Prediction of Future Land Use

The distribution of future land use types simulated by the GeoSOS-FLUS model is shown in Fig. 8.8.

As shown in Fig. 8.8, the area of farmland will be the largest in 2040, while the water will be more prominent in the northwest of the region. The proportion of land uses in 2040, calculated using ArcGIS, is presented in Table 8.7. Farmland will decrease by 4% compared with that in 2020, and the area of grassland (78 km^2) will decrease slightly. The proportions of remaining land use types will all increase, with the proportion of residential land reaching 17.75%.

According to the land use transfer matrix from 2020 to 2040 (Table 8.8), from 2020 to 2040, 0.85 km^2 of farmland will be converted to grassland, 56 km^2 to residential land, 0.18 km^2 to forest land, and 13.56 km^2 to water; 3.99 km^2 of grassland will be converted to farmland, only 0.07 km^2 to residential land, 0.01 km^2 to forest

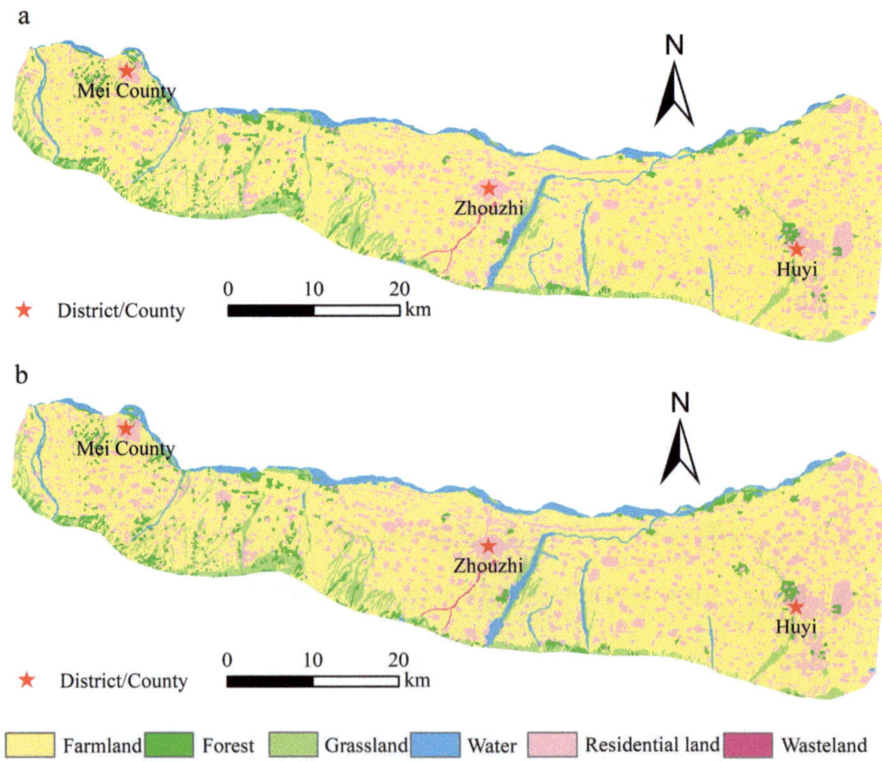

Fig. 8.8 Distribution of land use types. (**a**) Actual value in 2020, and (**b**) Predicted value in 2040

Table 8.7 Proportion of land use types in the study area in 2040

Type	Area (km²)	Proportion (%)
Farmland	1209.208	68.47
Forest	59.92	3.39
Grassland	102.996	5.83
Water	78.267	4.43
Residential land	313.459	17.75
Wasteland	2.22	0.13

Table 8.8 Land use transfer matrix from 2020 to 2040 (unit: km²)

	2040						
2020	Farmland	Forest	Grassland	Water	Residential land	Wasteland	Total
Farmland	1203.2	0.18	0.85	13.56	56	0.01	1273.8
Forest	0.68	59.7	0.015	0.035	0.04	0	60.47
Grassland	3.99	0.01	102.07	0.12	0.07	0	106.26
Water	0.05	0.013	0.04	64.54	0.007	0	64.65
Residential land	1.27	0.017	0.021	0.012	257.34	0	258.66
Wasteland	0.018	0	0	0	0.002	2.21	2.23
Total	1209.208	62.24	102.996	78.267	313.459	2.22	1766.07

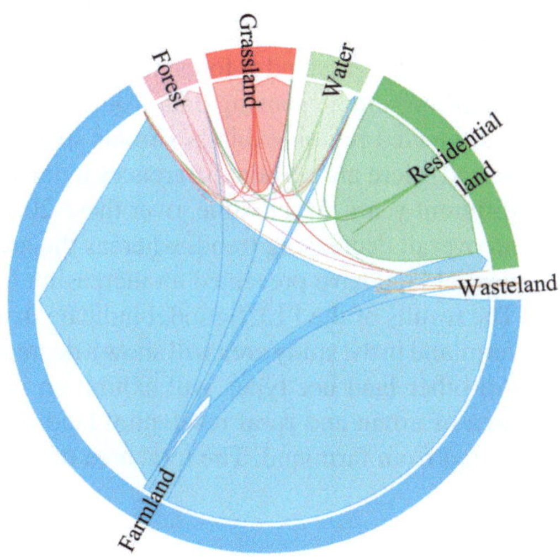

Fig. 8.9 Land use transfer chord map for 2020–2040

land, and 0.12 km² to water; 0.021 km² of residential land will be converted to grassland, 1.27 km² to farmland, 0.017 km² to forest land, and 0.012 km² to water area; 0.015 km² of forest will be converted to grassland, 0.04 km² to residential land, 0.68 km² to farmland, and 0.035 km² to water; 0.021 km² of water will be converted to grassland, 0.007 km² to residential land, 0.017 km² to forest, and 0.05 km² to farmland.

Overall, as shown in the chord diagram of land use transfer (Fig. 8.9), from 2020 to 2040, the total area of farmland will decrease, dropping to 1209.208 km², while the areas of other land types will all increase. Specifically, the area of residential land will increase by 54.8 km^2, and the area of water will increase by 13.6 km^2. The increased areas of residential land and water will be mainly converted from farmland.

8.5 Conclusions

This chapter elaborates on the characteristics of land use changes in the selected areas of the northern Qinling Mountains. Based on the remote sensing data of five periods (2000, 2005, 2010, 2015, and 2020), the land use in this selected area can be classified into six categories: farmland, forest, grassland, urban and rural residential land, water, and wasteland. The spatial distribution characteristics, structural changes, transfer processes, and spatial variations of land use types in the northern Qinling Mountains were investigated. Additionally, the distribution of land use types in 2040 was predicted using the GeoSOS-FLUS model. The following conclusions can be summarized:

1. Farmland constitutes the dominant land use type across the entire region. Grassland and forest are scattered in Mei County and the southern part of the region. Urban and rural residential land is interspersed within farmland areas, while wasteland is exclusively distributed in the central part of Zhouzhi County. According to the magnitude and dynamic degree of land use change, the relative intensity of land use are ranked as: water > urban and rural residential land > farmland > grassland > forest > wasteland.
2. Based on remote sensing data from 2000 to 2020, although the land use within the selected region exhibits similar spatial distribution characteristics over this period, there are certain differences in the area and proportion of land use across different years. Meanwhile, over these 20 years, the area of farmland has shown an overall decreasing trend, whereas the areas of water and urban and rural residential land have presented an increasing trend.
3. The results of the FLUS model indicate that from 2020 to 2040, the total area of farmland in the study area will show a decreasing trend, dropping to 1209.208 km^2. All other land use types will exhibit an increasing trend; specifically, the total area of urban and rural residential land will increase by 54.8 km^2, mainly converted from farmland. The total area of water will increase by 13.6 km^2.

References

Han XT, Bian ZF, Yu HC et al (2025) The historical and theoretical rationale for ecological protection and restoration: experiences from China. Land 14(1):161. https://doi.org/10.3390/land14010161

Lai JL, Qi S, Chen JD et al (2025) Exploring the spatiotemporal variation of carbon storage on Hainan Island and its driving factors: insights from InVEST, FLUS models, and machine learning. Ecol Indic 172:113236. https://doi.org/10.1016/j.ecolind.2025.113236

Li X, Chen G, Liu X et al (2017) A new global land-use and land-cover change product at a 1-km resolution for 2010 to 2100 based on human–environment interactions. Ann Am Assoc Geogr 107(5):1040–1059. https://doi.org/10.1080/24694452.2017.1303357

Li Y, Duan X, Li Y et al (2021) Interactive effects of land use and soil erosion on soil organic carbon in the dry-hot valley region of southern China. Catena 201:105187. https://doi.org/10.1016/j.catena.2021.105187

Li H, Zhang T, Cao XS et al (2022) Shaanxi Province (China) for ecological restoration of land space. Forests 13(5):766. https://doi.org/10.3390/f13050766

Li LX, Li PY, He S et al (2025) Groundwater nitrate pollution source apportionment under varying land use/land cover patterns. Expo Health 17:353–369. https://doi.org/10.1007/s12403-024-00666-0

Usman M, Liedli R, Shahid M et al (2015) Land use/land cover classification and its change detection using multi-temporal MODIS NDVI data. J Geogr 25(12):1479–1506. https://doi.org/10.1007/s11442-015-1247-y

Yang J, Xie BP, Zhang DG et al (2023) Habitat quality assessment and multi-scenario prediction of the Gansu–Qinghai section of the Yellow River Basin based on the FLUS-InVEST model. Front Ecol Evol 11:1228558. https://doi.org/10.3389/fevo.2023.1228558

Yang XL, Zhang JY, Yu YY et al (2024) Behavioral conflicts in urban greenway recreation: a case study of the "Three Rivers and One Mountain" Greenway in Xi'an, China. Land 13(7):1097. https://doi.org/10.3390/land13071097

Zhao R, Yang ZY, Zheng H et al (2022) A framework for the general design and computation of hybrid neural networks. Nat Commun 13:3427. https://doi.org/10.1038/s41467-022-30964-7

Zhu LY, Song RX, Sun S et al (2022) Land use/land cover change and its impact on ecosystem carbon storage in coastal areas of China from 1980 to 2050. Ecol Indic 142:109178. https://doi.org/10.1016/j.ecolind.2022.109178

Chapter 9
Groundwater Nitrate Pollution and Health Risks Under Land Use Changes

Abstract This chapter investigates the spatiotemporal distribution, sources, and health risks of groundwater nitrate pollution under land use changes in the selected area of the northern Qinling Mountains, focusing on Qishan–Mei–Zhouzhi–Huyi region. For this study, 64 groundwater samples were collected and analyzed to assess hydrochemical characteristics and nitrate concentrations, alongside the use of Piper diagrams to classify groundwater types. The Random Forest model identifies dominant factors influencing nitrate levels, while the Bayesian stable isotope mixing model (MixSIAR) quantifies the contributions of nitrate sources such as soil nitrogen, fertilizers, and sewage. Health risk assessment employs deterministic and stochastic (Monte Carlo simulation) approaches to evaluate non-carcinogenic risks for adults and children, with sensitivity analysis to address parameter uncertainties. The chapter highlights the interplay between land use patterns, particularly farmland and residential areas, and groundwater nitrate pollution, providing insights for sustainable land and water management. The findings highlight the need for targeted mitigation strategies to reduce health risks associated with nitrate contamination.

Keywords Groundwater pollution · Land use changes · Groundwater nitrate · Nitrate source identification · Health risk assessment

9.1 Introduction

Currently, groundwater nitrate pollution has become a widespread environmental problem worldwide (Vystavna et al. 2017). High concentrations of nitrate not only lead to eutrophication of rivers, but also cause problems such as methemoglobinemia or cancer in humans (Li et al. 2023; Dash et al. 2024). Based on this, it is important to study the spatiotemporal distribution, sources, influencing factors, and health risks of nitrate. Meanwhile, due to the advancement of urbanization and the rapid development of industry and agriculture, the types of land use have been changing. Existing studies have demonstrated that land use patterns are closely

P. Li et al., *Water Resources Research in Northern Qinling Mountains, China*, Springer Hydrogeology, https://doi.org/10.1007/978-3-032-09395-0_9

associated with groundwater nitrate content and also affects the amount of groundwater (He et al. 2020; Kaown et al. 2007; Sattolo et al. 2017; Wang et al. 2017). Wang et al. (2018) pointed out that in areas with frequent human activities, such as farmland and residential land, the concentration of groundwater nitrate is significantly greater than in areas with weaker human activities. Additionally, the implementation of land use planning has certain ecological effects on the surrounding environment. Over the years, significant changes have occurred in land use patterns in the northern Qinling Mountains, with a large amount of farmland converted into kiwifruit orchards. This is prevalent in the Mei County and Zhouzhi County of the study area, which have become the largest kiwifruit production bases in the world (Xie et al. 2024; Wang et al. 2025). However, the vigorous development of the kiwifruit industry, accompanied by prominent issues of irrational fertilizer application, has also brought about a series of environmental problems.

Based on Chap. 8, this chapter investigates the variations in groundwater nitrate levels from the perspective of land use change, further quantifies the various sources of nitrate, and evaluates the non-carcinogenic health risk of nitrate. Due to the numerous uncertain factors in the health risk evaluation model, the exposure parameters for populations in different areas will vary. Therefore, the parameters of the model are not a definite value and will be affected by many uncertainties such as body weight, water intake, and age. Therefore, to reduce errors generated by the parameters, this chapter also uses the Monte Carlo model to simulate the exposure parameters and risk characterization parameters, improving the accuracy of the research results.

9.2 Influence of Different Land Use Patterns on Groundwater Nitrate

9.2.1 Sample Collection and Analysis

In the field survey of the selected study area (Qishan County–Mei County–Zhouzhi County–Huyi District), a total of 64 samples of groundwater were collected, and the distribution map of groundwater sampling points is shown in Fig. 9.1. Three samples were taken from each water sampling point, two of them were placed in 500 mL sampling bottle and 2.5 L polyethylene plastic bucket for the determination of conventional physicochemical indices, and the other one was contained in 500 mL sampling bottle for the determination of isotope (^{15}N and ^{18}O) contents. All samples were collected, preserved and transported following national guidelines. The samples were sent to the Laboratory of Xi'an Mineral Resources Survey Center for K^+, Na^+, Ca^{2+}, Mg^{2+}, SO_4^{2-}, Cl^-, HCO_3^- and TDS indices; to Shaanxi Nuclear Industry Group Comprehensive Analytical Testing Co. Ltd. for TN, NO_3^-, NO_2^-, NH_4^+, Mn, and Fe indices; and the nitrogen–oxygen isotope indices were detected by Xi'an Mining Spectrum Geological Exploration Technology Co. Ltd. The field test

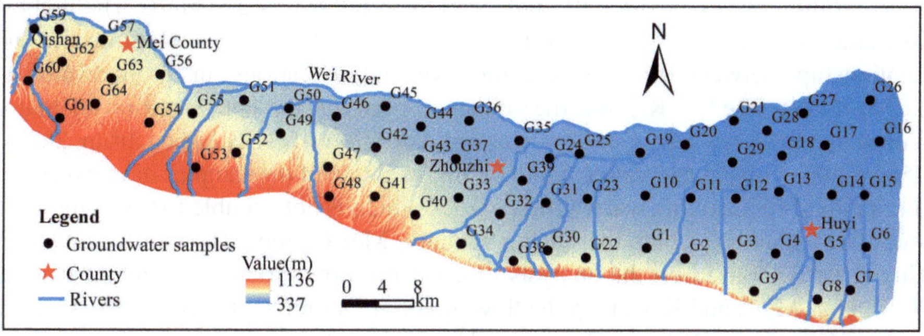

Fig. 9.1 Location distribution of groundwater sampling points in the selected study area

Table 9.1 Statistics of major ions and chemical indicators of groundwater in the study area (mg/L)

Areas	Statistics	Na^+	K^+	Ca^{2+}	Mg^{2+}	Cl^-	HCO_3^-	SO_4^{2-}	TDS	pH
Qishan County	Max	37.9	3.59	137	20	42.9	336	82.7	628	7.4
	Min	11.3	1.15	54.4	9.19	9.95	125	14	277	6.83
	Mean	20.23	2.23	83.35	14.42	19.61	219.5	41.13	403	7.16
Mei County	Max	40	3.59	205	49.1	96.1	373	103	1298	7.57
	Min	6.68	0.69	59.7	10.8	2.76	132	7.29	279	6.96
	Mean	18.99	2.10	122.8	25.44	27.68	257.43	59.36	747.57	7.29
Zhouzhi County	Max	69	4.03	195	44.7	87.1	380	183	1121	7.86
	Min	5.04	0.48	61.2	10.2	2.6	153	5.59	262	7.21
	Mean	20.62	2.15	104.63	22.03	19.97	253.5	59.05	516.04	7.49
Huyi District	Max	308	4.24	220	161	210	642	997	2566	7.81
	Min	4.83	0.64	70.9	14.2	5.49	162	36.1	300	7.16
	Mean	54.29	2.27	142.99	43.12	46.52	348.67	188.19	832.89	7.47

indicators include: water temperature (T), pH, electrical conductivity (EC), dissolved oxygen (DO), and oxidation–reduction potential (ORP).

9.2.2 *Basic Hydrochemical Characteristics of Groundwater*

The statistical results of the main chemical indices of groundwater in the selected study area are shown in Table 9.1. The pH value of groundwater in the area ranged from 6.83 to 7.86, showing an overall neutral to slightly alkaline nature. The concentration of Na^+ varies from 4.83 to 308 mg/L, while the content of K^+ was the lowest in groundwater, with the concentration ranging from 0.48 to 4.24 mg/L. The concentration ranges of Ca^{2+} and Mg^{2+} were 54.40 to 220 mg/L and 9.19 mg/L to 161 mg/L, respectively, with the mean values of 118.06 mg/L and 28.23 mg/L, respectively. The major anion with the largest range of variation was HCO_3^-, which ranged from 125 to 642 mg/L, with a mean value of 279 mg/L. The lowest concentration of SO_4^{2-} was 5.59 mg/L, and the highest was 279 mg/L. The highest concentration was 997 mg/L, and the average concentration was 94.32 mg/L. The relationship between the concentration of major cations in groundwater is $Ca^{2+} > Na^+ > Mg^{2+} > K^+$, and the sequence between the concentration of major anions is $HCO_3^- > SO_4^{2-} > Cl^-$. The concentration of TDS varied from 262 to 2566 mg/L, with a mean value of 648.73 mg/L. Some of the samples have exceeded the standard for drinking water (1000 mg/L), which is not suitable for drinking. The study area mainly includs the Qishan County, Mei County, Zhouzhi County, and Huyi District. Based on the analysis of concentration variations of major cations (Ca^{2+}, Na^+, Mg^{2+}, and K^+), along the flow direction of the Wei River (Qishan County → Mei County → Zhouzhi County → Huyi District), the concentrations of Ca^{2+}, Mg^{2+}, and K^+ trend to be the highest in the Huyi District. Na^+ was the lowest in the Huyi District, while K^+ was the lowest in Mei County. Additionally, the concentrations of Ca^{2+} and Mg^{2+} were the lowest in Qishan County. According to the analysis of the concentration variations of major anions (HCO_3^-, SO_4^{2-}, Cl^-), it is observed that HCO_3^-, SO_4^{2-} and Cl^- all follow the pattern of having the highest concentration in the Huyi District, while the concentrations of SO_4^{2-}, Cl^- and HCO_3^- are the lowest in the Qishan County.

A total of 4 groundwater samples were collected in the Qishan County, with pH values ranging from 6.83 to 7.40, exhibiting slightly alkaline properties. The concentration of Na^+ varies from 11.30 to 37.90 mg/L, while K^+ has the lowest content in the groundwater, with a concentration range of 1.15 to 3.59 mg/L. This is primarily attributed to the weak mobility of K^+, which is determined by its biological activity. The concentration ranges of Ca^{2+} and Mg^{2+} are 54.40–137 mg/L and 9.19–20 mg/L, with average values of 83.35 mg/L and 14.42 mg/L, respectively. Among the major anions, Cl^- shows the smallest concentration variation range, and HCO_3^- exhibits the largest, with their concentration ranges being 9.95–42.90 mg/L and 125–336 mg/L, respectively. The concentration of SO_4^{2-} ranges from 14 to 82.70 mg/L, with an average of 41.13 mg/L. In the groundwater of the Qishan County, the concentration order of major anions is $HCO_3^- > SO_4^{2-} > Cl^-$, and that of major cations is $Ca^{2+} > Na^+ > Mg^{2+} > K^+$. The TDS concentration ranges from a minimum of 277 mg/L to a maximum of 628 mg/L, with an average of 403 mg/L, which meets the quality standards for drinking water.

A total of 14 groundwater samples were collected in the Mei County. The pH of the groundwater ranges from 6.96 to 7.57, showing a slightly alkaline nature. Among the ions in the groundwater, K^+ has the lowest content, with a mean concentration of 2.10 mg/L. The concentration of Na^+ varies from 6.68 to 40 mg/L, while the concentration ranges of Ca^{2+} and Mg^{2+} are 59.7–205 mg/L and 10.8–49.1 mg/L, with mean values of 122.80 mg/L and 25.44 mg/L, respectively. Among the major anions, the concentration ranges of Cl^- and HCO_3^- are 2.76–96.10 mg/L and 132–373 mg/L, respectively. The concentration of SO_4^{2-} ranges from 7.29 to

103 mg/L, with an average of 59.36 mg/L. In the groundwater, the order of major cation concentrations is $Ca^{2+} > Mg^{2+} > Na^+ > K^+$, and the order of major anion concentrations is $HCO_3^- > SO_4^{2-} > Cl^-$. The concentration of TDS ranges from 279 to 1298 mg/L, with an average of 747.57 mg/L, and some samples do not meet the quality standards for drinking water.

A total of 28 groundwater samples were collected in the Zhouzhi County. The pH value ranges from 7.21 to 7.86. The concentration of Na^+ varies from 5.04 to 69 mg/L, while the range of K^+ concentration is 0.48 to 4.03 mg/L. The concentration ranges of Ca^{2+} and Mg^{2+} are 61.2–195 mg/L and 10.2–44.7 mg/L, with average values of 104.63 mg/L and 22.03 mg/L, respectively. Among the major anions, the concentration ranges of HCO_3^- and SO_4^{2-} are 153–380 mg/L and 5.59–183 mg/L, respectively. For Cl^-, the minimum concentration is 2.6 mg/L, the maximum is 87.1 mg/L, and the average concentration is 19.97 mg/L. The concentration order of the major anions in groundwater is $HCO_3^- > SO_4^{2-} > Cl^-$, and that of the major cations is $Ca^{2+} > Mg^{2+} > Na^+ > K^+$. The concentration of TDS ranges from 262 to 1121 mg/L, with an average of 516.04 mg/L, which basically meets the quality standards for drinking water.

A total of 18 groundwater samples were collected in the Huyi District. The pH value ranges from 7.16 to 7.81, showing a slightly alkaline property. The range of K^+ concentration is the smallest, varying from 0.64 to 4.24 mg/L, while the concentration of Na^+ ranges from 4.83 to 308 mg/L. The concentration ranges of Ca^{2+} and Mg^{2+} are 70.9–220 mg/L and 14.2–161 mg/L, with average values of 142.99 mg/L and 43.12 mg/L, respectively. Among the major anions, the minimum concentration of Cl^- is 5.49 mg/L, the maximum is 210 mg/L, and the average concentration is 46.52 mg/L. The concentration ranges of HCO_3^- and SO_4^{2-} are 162–642 mg/L and 36.1–997 mg/L, respectively. The concentration order of the major anions in groundwater is $HCO_3^- > SO_4^{2-} > Cl^-$, and that of the major cations is $Ca^{2+} > Na^+ > Mg^{2+} > K^+$. The concentration of TDS ranges from 300 to 2566 mg/L, with a mean of 832.89 mg/L, and some samples do not meet the quality standards for drinking water.

Based on the water quality monitoring results of groundwater sampling points, the major anions and cations in the samples were analyzed, and a Piper trilinear diagram of the hydrochemical types of groundwater was plotted (Fig. 9.2). As shown in Fig. 9.2, the groundwater in the Qishan County and Mei County is of the HCO_3–Ca·Mg type, while that in the Zhouzhi County and Huyi District is predominantly of the HCO_3–Ca·Mg type, with a small number of samples belonging to the SO_4·Cl–Ca·Mg type. In terms of the dominant anions and cations, Ca^{2+} is the dominant cation in the groundwater of the Qishan County, Mei County, Zhouzhi County, and Huyi District. HCO_3^- is the dominant anion in all groundwater samples within the study area. Only a few individual groundwater samples in the Huyi District exhibit a tendency for SO_4^{2-} to become the dominant ion.

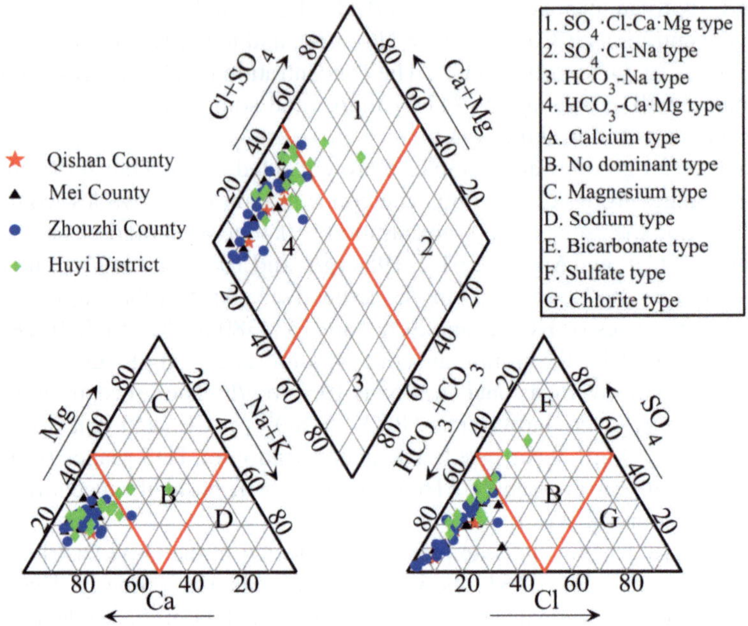

Fig. 9.2 Piper trilinear diagram of groundwater hydrochemical types

9.2.3 Groundwater Nitrate Spatial Distribution

In this study, regions where the nitrate concentration in groundwater exceeds 50 mg/L are defined as exceeding the standard, based on the nitrate limit for drinking water specified by the WHO. The statistical results of the distribution of groundwater nitrate in each district and county within the study area are presented in Table 9.2.

The concentration of nitrate in groundwater within the study area is 89.99 ± 1.42 mg/L, with the mean value significantly exceeding the limit (50 mg/L). The maximum concentration reaches 397 mg/L, and the average over-standard rate among 64 samples is 55%, indicating that groundwater in the area is threatened by nitrate pollution. As shown in Table 9.2, the nitrate concentration is the highest in the Mei County with a relatively large variation range of 4.5–397 mg/L. The concentration of NO_3^- in the Zhouzhi County ranges from 0 to 287 mg/L. The Qishan County has the lowest NO_3^- concentration, with an average of 46.33 mg/L. The average concentration of NO_3^- in groundwater in the Huyi District is 66.82 mg/L. The spatial distribution map of groundwater nitrate concentration (Fig. 9.3) was obtained through spatial interpolation analysis.

In the horizontal direction, along the flow direction of the Wei River (Qishan County → Mei County → Zhouzhi County → Huyi District), the spatial distribution of nitrate in the study area is uneven. The sampling points with relatively higher concentrations (G18, G44, G63, G64) are located in the Laodian Street of the Huyi District, the Yabai Town of the Zhouzhi County, and the Qi Town of the Mei County, respectively. Moreover, the average NO_3^- concentration near the highest value point

Table 9.2 Nitrate distribution statistics of groundwater in the study area

Item	Qishan County	Mei County	Zhouzhi County	Huyi District
Max (mg/L)	67	397	287	313
Min (mg/L)	30.8	4.5	0	0
Mean (mg/L)	46.33	160.88	75.68	66.82
Standard deviation (mg/L)	4.41	8.36	2.55	4.78
Sample over standard rate (%)	100	92.9	85.7	66.7

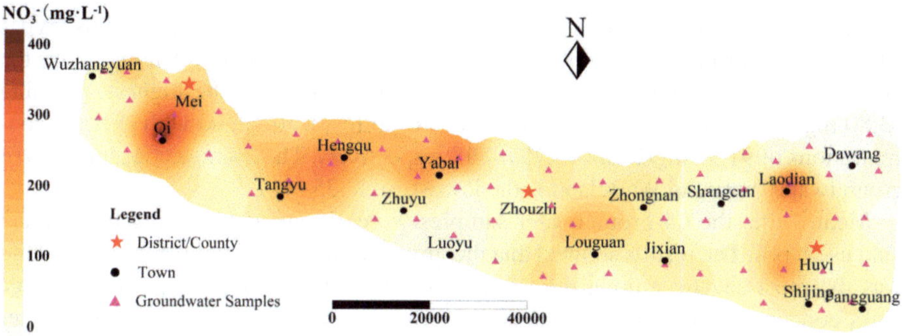

Fig. 9.3 Spatial distribution of nitrate (NO_3^-) concentration in the study area

exceeds 100 mg/L, and surrounding land use types are mainly farmland and residential land dominated by rural settlements. In the southern part of the study area near the Qinling Mountains and the northern part of the Dawang Street, the groundwater NO_3^- concentration is relatively low, with the surrounding land use types being grassland and urban residential areas. Therefore, in terms of groundwater nitrate pollution across the entire region, the nitrate concentration in more than half of the area exceeds Class V, and only the southeast corner of the Zhuyv Town, Shangcun Town, and the vicinity of the Dawang Street have nitrate concentrations better than Class III. The reasons for the above distribution characteristics may be that the soil and aquifers in the study area have good permeability. A large amount of domestic sewage discharge, agricultural irrigation, and chemical fertilizer application from dense urban–rural residential areas, agricultural areas, and industrial parks lead to nitrate infiltration into the soil and migration to aquifers, resulting in high nitrate concentrations in groundwater.

9.2.4 *Groundwater Nitrate Distribution Under Land Use Changes*

Based on ArcGIS, the content of groundwater nitrate under different land use types in the study area in 2020 was statistically analyzed (Table 9.3). The order of groundwater nitrate concentrations across different land use types is as follows: Farmland > Forest ≈ Urban and Rural residential land > Grassland > Water > Wasteland. Except for

Table 9.3 Statistical values of nitrate under different land use types (mg/L)

Land use type	Max	Min	Mean	Standard deviation
Farmland	380	0	93.80	69.09
Forest	320	0	91.99	67.73
Grassland	292	0	72.80	59.96
Water	221	0	63.30	46.13
Urban and rural residential land	380	0	91.75	67.49
Wasteland	52	7	28.35	11.83

Wasteland, the average concentrations of groundwater nitrate in other land use types all exceed the drinking water nitrate limit specified by the WHO. Among them, the average concentration of groundwater nitrate in Farmland is the highest, reaching 93.80 mg/L, followed by Forest and Urban-Rural residential land with average concentrations of 91.99 mg/L and 91.75 mg/L, respectively. The average concentration of groundwater nitrate in Wasteland is the lowest, at 28.35 mg/L.

Meanwhile, the 64 groundwater samples were classified according to different land use types. The results showed that the groundwater nitrate pollution in Farmland was the most severe: among the 38 samples from this land use type, the average mass concentration of nitrate was 88.2 mg/L, which exceeded the drinking water nitrate limit specified by the WHO, with an over-standard rate of 50%. Existing studies have found that only 30–40% of nitrogen fertilizers used in China can be absorbed and utilized by crops, while the rest enter groundwater through leaching via irrigation and rainfall or remain in the soil (Xin et al. 2019). Therefore, the irrational use of nitrogen-containing fertilizers allows nitrogen to easily enter groundwater in the form of nitrate. Additionally, the use of sewage for farmland irrigation also causes nitrate pollution in groundwater within agricultural land. These factors contribute to the high nitrate concentration in groundwater of Farmland, thereby leading to groundwater pollution. Secondly, the sample test results indicated that the nitrate concentration in Urban and Rural residential land was 83.1 mg/L, second only to that in Farmland. This may be attributed to the rapid development of urban and rural areas and high population density, especially the construction of tourism industries and industrial parks, which have increased sewage discharge in various regions, thereby resulting in severe nitrate pollution in urban–rural groundwater. Furthermore, most groundwater sampling points in residential land are mostly in rural settlements, where nitrate concentrations are concentrated. This may be because domestic sewage in rural settlements is not effectively collected and managed in a unified manner, and the impermeable area in this region is small, making infiltration more likely to cause the above results.

9.2.5 Nitrate Levels Under Future Land Use Change

In 2020, the primary land use types within the study area were Farmland and Urban-Rural land. The future distribution of land use types simulated using the GeoSOS-FLUS model is shown in Fig. 8.7. As indicated in Fig. 8.7, Farmland will remain the largest land use type in terms of area by 2040, although its area will decrease, while the area of Water will expand.

According to the land use transfer matrix from 2020 to 2040 (Table 8.8), the area of Farmland will decrease, primarily converting to Urban and Rural residential land with a converted area of 54.8 km^2. Secondly, the area of Water will increase relatively by 13.6 km^2, mainly converted from Farmland. Considering the current pollution mechanisms and future land use changes, Farmland and Urban-Rural residential land will still dominate the region in 2040. Due to factors such as agricultural fertilization, irrigation, and domestic sewage discharge, the nitrate concentration in groundwater within the study area will remain at a relatively high level. Moreover, due to the slow renewal rate of groundwater circulation, the water quality of polluted groundwater will be difficult to restore within a certain period. Additionally, the increased area of water will be mainly distributed in the region north of the Mei County, where the nitrate concentration in groundwater may decrease.

Therefore, in future economic development, domestic and industrial sewage should be discharged according to standard, and strict measures should be taken to prevent leakage during the sewage discharge process. Agricultural fertilization patterns in farmland need to be improved (Li et al. 2025). For instance, low-concentration compound fertilizers suitable for specific crops should be formulated by combining organic compost with inorganic chemical fertilizers; the usage of chemical fertilizers should be reduced, especially in regions with high fertilizer demand such as the Mei County and Zhouzhi County, thereby mitigating groundwater nitrate pollution. Rural residential areas should adopt centralized domestic wastewater treatment as much as possible and select better drinking water sources to avoid a series of health problems caused by long-term consumption of groundwater with excessive pollutants.

9.3 Main Controlling Factors and Sources of Groundwater Nitrate

9.3.1 Research Methods

9.3.1.1 Random Forest Model

Random Forest (RF) analysis is an algorithm that optimizes predictions based on multiple decision trees and falls under the category of machine learning algorithms (Fig. 9.4). It is a classifier composed of multiple decision trees. The classification results are determined by evaluating the attributes of test samples across various dimensions on different decision trees (He et al. 2022). After comprehensively considering all the individual decisions, the final classification is derived. For classification problems, the class with the highest probability is selected as the result, while for regression analysis, the average of the predicted values is taken. This approach enables efficient and rapid identification of the most important feature categories for sample classification.

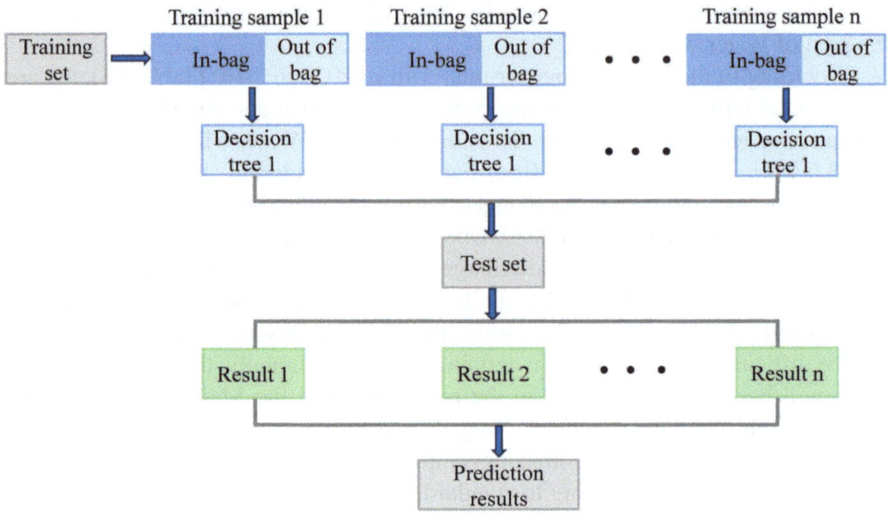

Fig. 9.4 Schematic diagram of the principle of random forests

RF employs the "bootstrap" algorithm to randomly sample from the training dataset and generate decision trees. These trees are then combined through a method called "bagging" (bootstrap aggregating) to form the forest. The initial training dataset is split into two parts: "in-bag" and "out-of-bag" samples. Observations that exist in the training dataset but are not included in the randomly selected subset used for training each individual tree are referred to as "out-of-bag" samples, which are used to evaluate the model's performance. In contrast, "in-bag" samples are used to construct each decision tree within the forest, while "out-of-bag" samples allow for internal validation. The bootstrap algorithm generates numerous new samples of the same size directly from the original dataset without relying on external data sources. This method is particularly valuable when dealing with small sample sizes. Traditional training-test splits on small datasets can lead to increased bias. The bootstrap approach maintains the original training sample size while still providing a validation set (because the training set of repetitive, but this repetition is random), thus offering certain advantages.

RF algorithm eliminates generalization errors through a large number of decision trees, thereby improving prediction accuracy (Tesoriero et al. 2017). Since RF can well describe nonlinear relationships, the importance of each dependent variable to classification events can be obtained by establishing a Random Forest Classification (RFC) model to classify dependent variables. In recent years, RF algorithm has been regarded as an effective method to identify the dominant factors of pollutants present in groundwater (Rodriguez-Galiano et al. 2014; Podgorski et al. 2018; Wei et al. 2021).

9.3.1.2 Bayesian Stable Isotope Mixing Model

The Bayesian stable isotope mixing model (MixSIAR) uses Monte Carlo–Markov Chain (MCMC) simulations within a Bayesian framework to calculate the contribution ratios of various sources to the sample mixture as indicated by isotopic tracers. Stock et al. (2018) proposed the use of the Bayesian stable isotope mixing model to estimate the proportion of nitrate (NO_3^-) sources, and this process is implemented through the MixSIAR package (version 3.1.12) created and run in R.

Since MixSIAR is a logical prior distribution constructed within the Bayesian framework based on the Dirichlet distribution, the model can be expressed as Eq. (9.1) (Li et al. 2024):

$$\begin{aligned} X_{ij} &= \sum\nolimits_{k=1}^{k} p_{ik}\left(S_{jk} + C_{jk}\right) + \varepsilon_{ij} \\ S_{jk} &\sim N\left(\mu_{jk}, \omega_{jk}^{2}\right) \\ C_{jk} &\sim N\left(\lambda_{jk}, \tau_{jk}^{2}\right) \\ \varepsilon_{ij} &\sim N\left(\mathbf{0}, \sigma_{ij}^{2}\right) \end{aligned} \tag{9.1}$$

where, X_{ij} represents the jth isotopic composition in the ith sample; p_{ik} is the contribution of kth source to the ith sample; S_{jk} denotes jth isotopic value in the kth source, which present a normal distribution with an average μ_{jk} and variance $\omega^2{}_{jk}$; C_{jk} denotes the fractionation factor of isotope j on source k; λ_{jk} and $\tau^2{}_{jk}$ are the average and variance of the fractionation factors, respectively; ε_{ij} is the residual error, presenting a normal distribution with an average of zero and variance of $\sigma^2{}_{ij}$.

9.3.2 *Main Control Factors of Groundwater Nitrate Concentration*

9.3.2.1 Model Construction and Testing

The concentration distribution of nitrate components in groundwater is primarily controlled by their sources, reaction processes, and migration pathways. In this section, the Random Forest model was employed to identify parameters with high importance for nitrate (NO_3^-), analyze the relationships between these parameters and NO_3^-, and determine the main factors controlling the concentration distribution of nitrate pollutants. Accordingly, a Random Forest Classification model was developed based on data from collected groundwater samples, incorporating 17 predictor variables: K^+, Na^+, Ca^{2+}, Mg^{2+}, SO_4^{2-}, Cl^-, HCO_3^-, TDS, TN, NO_2^-, Fe, Mn, EC, Eh, pH, DO, T. The model was constructed using the Random Forest module in Python,

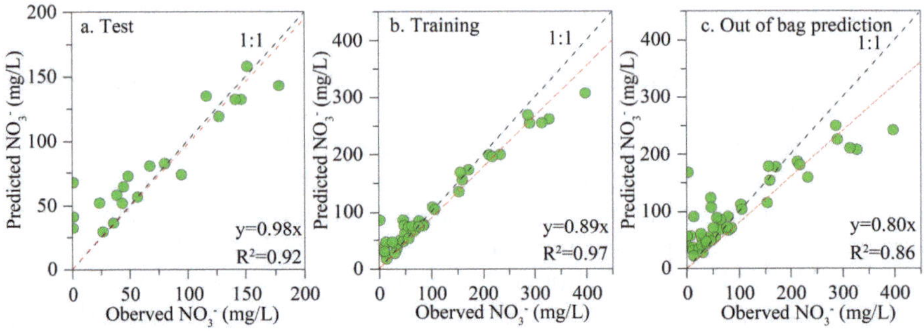

Fig. 9.5 Relationship between the predicted and measured values of the random forest model of groundwater nitrate value

with 70% of the data randomly selected as the training set and the remaining 30% used for validation. The relationships between predicted and measured values of NO_3^- for the training dataset, test dataset, and "out-of-bag" samples are shown in Fig. 9.5. RF algorithm combines "bagging" with random variable selection at each split. During bagging process, multiple decision trees are generated from bootstrap samples of the training data. Each tree produces a prediction, and the average of these predictions is computed to yield the optimal forecast.

For the training dataset, the coefficient of determination (R^2) between predicted and measured NO_3^- values was 0.97. For the test dataset, the R^2 was 0.92, and for "out-of-bag" predictions, the R^2 was 0.86. These results indicate that the Random Forest model for groundwater nitrate exhibits good predictive performance and can be used for further analysis.

9.3.2.2 Modelling Results

The results of the relative importance of predictor variables for groundwater nitrate values calculated by the random forest model are shown in Fig. 9.6. The random forest model results indicate that the variables with the greatest influence on NO_3^- in groundwater are, in order, TN, TDS, K^+, Ca^{2+}, and EC. Among these, the concentrations of TN, K^+, and TDS indicate that high concentrations of NO_3^- are related to agricultural activities and natural nitrogen cycle processes, such as fertilization and irrigation. The land use in the Zhouzhi County and Mei County at the northern Qinling Mountains is dominated by kiwifruit orchards, which have become the largest contiguous kiwifruit planting area in China and even the world. Due to excessive fertilization, the annual nitrogen surplus in kiwifruit orchards averages 1260 kgN/hm^2, and the nitrogen use efficiency is less than 20%. The accumulated nitrate nitrogen in the 0–10 m soil profile of kiwifruit orchards averages as high as 7113 kgN/hm^2, which is 8 times that of cultivated land soil. It is estimated that the accumulated nitrate nitrogen in the 10 m depth of the unsaturated zone in the kiwifruit producing areas of the Zhouzhi and Mei Counties reaches 266.5 Gg N (Gao et al. 2021). This

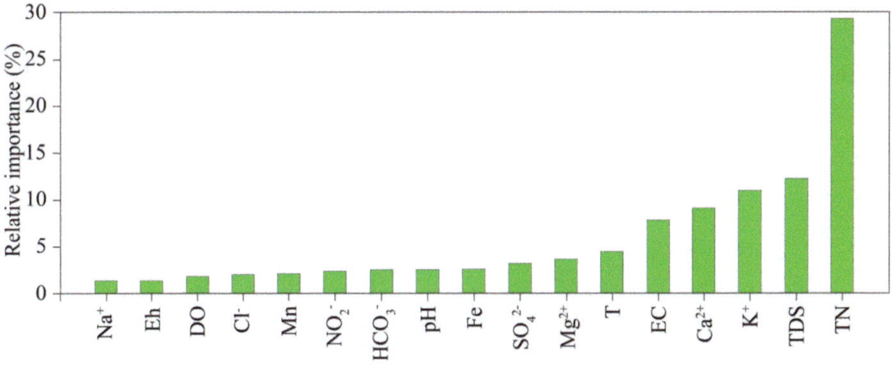

Fig. 9.6 Groundwater nitrate concentration and variable factors relative to significance

indicates that agricultural activities in kiwifruit orchards in this region have led to the accumulation of a large amount of NO_3^--N in the unsaturated zone. In addition, irrigation in orchards causes the nitrate nitrogen accumulated in the unsaturated zone to migrate into groundwater with water, posing a serious impact on groundwater quality.

9.3.3 Sources and Contributions of Groundwater Nitrate

In the process of groundwater movement, chloride ions (Cl^-) exhibit stable properties and generally do not undergo significant changes, thus often serving as tracers for identifying sources of nitrate (NO_3^-) pollution. By analyzing the relationship between NO_3^-/Cl^- and Cl^- in regional groundwater, the sources of NO_3^- in groundwater can be discriminated. The relationship between NO_3^-/Cl^- and Cl^- is illustrated in Fig. 9.7a, which reflects five input sources of nitrate in groundwater, including atmospheric deposition (AD), agricultural input (AI), soil nitrogen input (SN), animal manure and sewage (M&S). Sampling points in the study area with nitrate concentrations ranging from 0 to 88.57 mg/L mainly fall within the region corresponding to soil nitrogen. In contrast, water sample points with nitrate concentrations exceeding 88.57 mg/L (the limit value for Class III water in China's Groundwater Quality Standard), i.e., those with excessive nitrate concentrations, primarily lie within the wastewater region. This indicates that the main sources of groundwater nitrate in the study area are wastewater and leaching input from agricultural soil nitrogen. In addition, the relationship between dual isotopes of nitrate ($\delta^{15}N$-NO_3^- and $\delta^{18}O$-NO_3^-) is also commonly used to identify potential sources of groundwater nitrate. Statistical analysis shows that among the sampling points for isotopes measured in regional groundwater, the $\delta^{15}N$-NO_3^- values range from 1.17 to 10.66‰, with a mean of 4.89‰. The $\delta^{18}O$-NO_3^- values range from −4.53‰ to 5.16‰, with a mean of −1.92‰. The dual isotope relationship diagram of nitrate

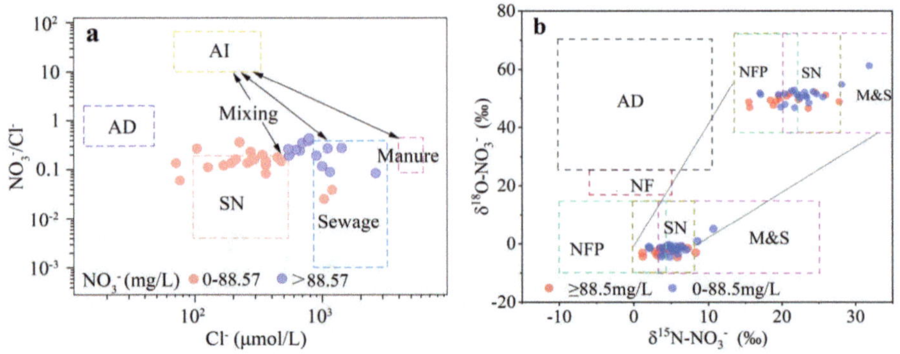

Fig. 9.7 Sources of nitrate based on hydrochemistry and isotopes. (**a**) NO_3^-/Cl^- vs. Cl^- and (**b**) relationship between $\delta^{15}N\text{-}NO_3^-$ vs $\delta^{18}O\text{-}NO_3^-$

(Fig. 9.7b) classifies nitrate into five regions based on the $\delta^{18}O\text{-}NO_3^-$ and $\delta^{15}N\text{-}NO_3^-$ characteristics of potential nitrate sources, which is roughly consistent with the classification in the NO_3^-/Cl^- vs. Cl^- diagram. Among these regions, there are overlapping areas between the zones corresponding to NH_4^+ input from precipitation and fertilizers (NFP), soil nitrogen input (SN), animal manure and sewage input (M&S). Most sampling points fall into these overlapping areas, indicating that the sources of nitrate in groundwater include NFP, SN, and M&S.

One of the most critical water environmental characteristics influencing groundwater nitrate contamination is dissolved oxygen (DO). Studies by Rivett et al. (2008) and Jahangir et al. (2013) have revealed that denitrification, which causes isotopic fractionation, may occur when the mass concentration of DO is below 2 mg/L, whereas denitrification does not take place if the DO content exceeds 2 mg/L. The DO values of the sampling points in the study area range from 2.30 to 9.11 mg/L, with a mean value of 6.10 mg/L. As shown in Fig. 9.8a, the DO values of all sampling points for nitrogen and oxygen isotopes are greater than 2 mg/L, failing to meet the conditions for denitrification to occur. In addition, denitrification and nitrification are primarily oxidation–reduction reactions, and the Eh exerts a significant impact on the functionality of these reactions. Jahangir et al. (2013) also found that an oxidation–reduction potential below 100 mV is conducive to the occurrence of denitrification. It can be seen from Fig. 9.8b that the Eh values of the sampling points range from 121 to 533 mV, with a mean value of 200 mV.

Denitrification leads to significant enrichment of nitrogen and oxygen stable isotopes. The occurrence of denitrification in groundwater can be determined based on the linear negative correlation between $\delta^{15}N\text{-}NO_3^-$, $\delta^{18}O\text{-}NO_3^-$ and NO_3^-. Relationship diagrams of groundwater $\delta^{15}N\text{-}NO_3^-$ versus NO_3^- and $\delta^{18}O\text{-}NO_3^-$ versus NO_3^- were plotted (Fig. 9.9a, c). Scatter fitting showed a negative correlation with poor correlation, further indicating that the groundwater is not conducive to denitrification. In addition, the relationship between ln (NO_3^-/Cl^-) and $\delta^{15}N\text{-}NO_3^-$, $\delta^{18}O\text{-}NO_3^-$ in the groundwater of the study area are very low, with correlation coefficients of 0.381 and 0.180, respectively. This also indicates that the nitrogen and

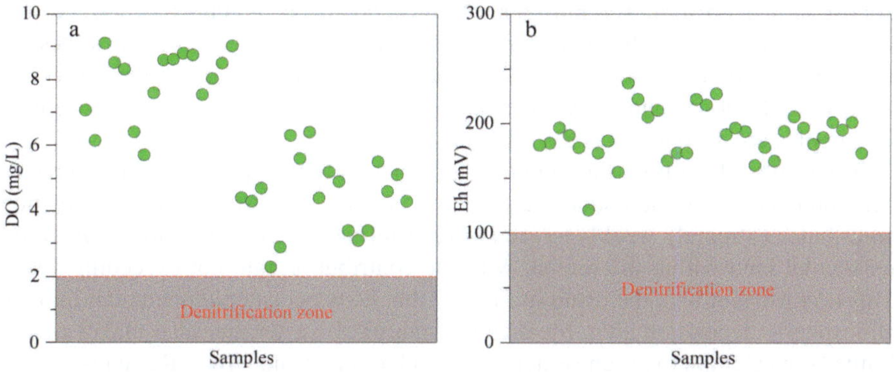

Fig. 9.8 Distribution of DO concentration at sampling points in the study area (**a**); Distribution of redox potential in the study area (**b**)

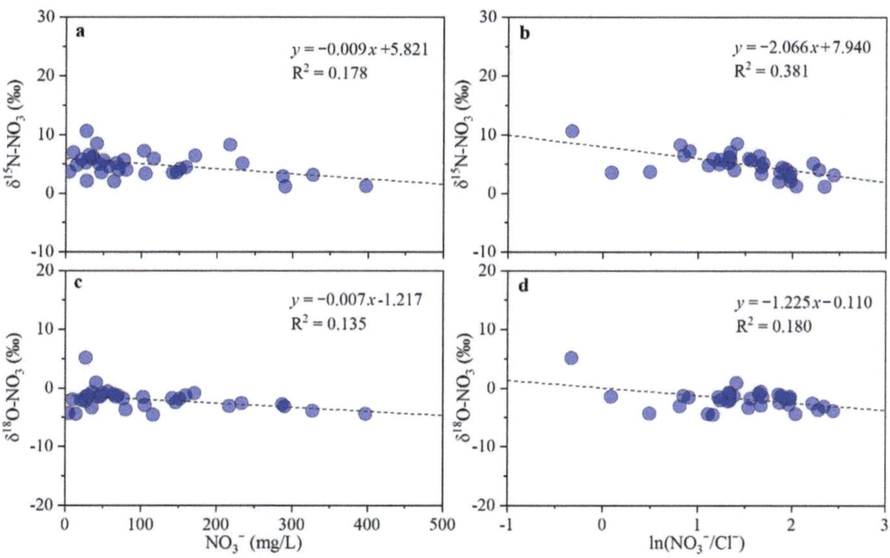

Fig. 9.9 Indicator relationship plots: (**a**) $\delta^{15}N\text{-}NO_3$ and NO_3^-; (**b**) $\delta^{15}N\text{-}NO_3$ and $\ln(NO_3^-/Cl^-)$; (**c**) $\delta^{18}O\text{-}NO_3$ and NO_3; (**d**) $\delta^{18}O\text{-}NO_3$ and $\ln(NO_3^-/Cl^-)$

oxygen isotopes of groundwater nitrate in the region are minimally affected by denitrification.

Statistical analysis reveals that the sampling points for nitrogen and oxygen isotopes are primarily distributed in land use types of Farmland and Urban-Rural residential land. Therefore, Farmland and Urban-Rural residential land are taken into account when quantitatively calculating the sources of groundwater nitrate. With reference to existing literature and considering the specific conditions of the study area, the following potential nitrate sources are identified in this research:

atmospheric deposition, soil nitrogen, chemical fertilizers, manure and sewage, and industrial wastewater. The characteristics of nitrogen and oxygen dual isotopes for each source are presented in Table 9.4. These five sources are set as the sources in the MixSIAR model, and the nitrogen and oxygen dual isotope values of groundwater samples are designated as the mixtures in the model.

The results of the Bayesian stable isotope mixing model quantifying the percentage contribution of each source are shown in Fig. 9.10. Atmospheric deposition contributes extremely weakly to groundwater nitrate in Farmland and Urban-Rural residential land within the region, with the contribution rate not exceeding 1.0%. This is consistent with the result in Fig. 9.7 that no water sample points fall into the atmospheric deposition area. Meanwhile, industrial wastewater also makes a very limited contribution to groundwater nitrate in Farmland and Urban-Rural residential areas, with contribution rates of 0.4% and 1.1% respectively, indicating that

Table 9.4 $\delta^{15}N$, $\delta^{18}O$ for potential sources of groundwater NO_3-N in the MixSIAR model (unit: ‰)

	$\delta^{15}N$-NO_3		$\delta^{18}O$-NO_3	
Source	Mean	Standard deviation	Mean	Standard deviation
Atmospheric deposition	0.10	1.70	55.00	7.60
Soil nitrogen	3.74	2.09	2.39	2.24
Chemical fertilizer	1.46	4.74	3.44	2.47
Manure and sewage	16.30	5.70	7.00	2.70
Industrial wastewater	8.50	0.90	38.50	18.90

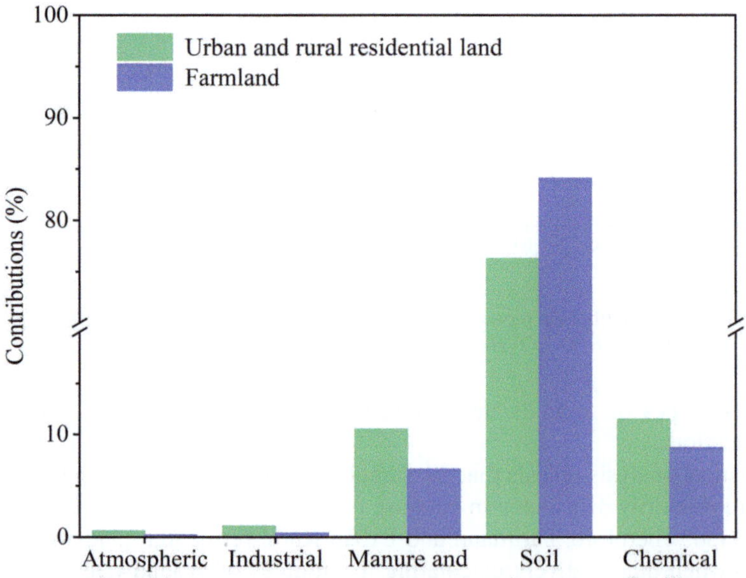

Fig. 9.10 Contribution ratio of groundwater nitrate sources corresponding to different land use types

groundwater nitrate in the study area is barely polluted by industrial wastewater. Soil nitrogen has the highest contribution to groundwater nitrate in Farmland and Urban and Rural residential land in the region, accounting for 84.1% and 76.3% respectively, followed by chemical fertilizers, manure and sewage. This suggests that soil nitrogen, chemical fertilizers, manure, and sewage are the main sources of groundwater nitrate. The reason why soil nitrogen and chemical fertilizers have become the primary sources of groundwater nitrate in Farmland and Urban-Rural residential land of the study area is that kiwifruit cultivation has been carried out in the Zhouzhi County and Mei County at the northern Qinling Mountains since 1990. The land use pattern has shifted from being dominated by cropland for growing wheat and corn to being dominated by kiwifruit orchards. Improper fertilization in kiwifruit production leads to low plant absorption rates and a large accumulation of soil nitrogen. The accumulated soil nitrogen infiltrates and leaches into groundwater through precipitation and irrigation water, thereby resulting in high nitrate concentrations in groundwater.

9.4 Groundwater Nitrate Health Risk Assessment

9.4.1 Research Methods

9.4.1.1 Deterministic Risk Assessment

The health risk assessment model adopts the "4-step method" recommended by the United States Environmental Protection Agency (USEPA): hazard identification, dose-effect analysis, exposure assessment, and risk characterization (Li et al. 2019; Fida et al. 2023; Guo et al. 2024).

The first step is hazard identification. Based on the background of widespread nitrate exceeding the standard in the study area, pollutants in drinking water that may pose potential health risks to the population are identified, and the hazards caused by these pollutants to human health are determined.

The second step is dose-effect analysis. By analyzing the dose-effect relationship, the extent of adverse health effects on the population caused by the exposure concentration of a specific pollutant under certain exposure conditions can be obtained. Depending on the duration of the response, dose-effect relationships can be classified into three types: acute, subchronic, and chronic responses. Through dose-effect analysis, pollutants in groundwater can be categorized into non-carcinogenic substances and carcinogenic substances.

The third step is exposure assessment, which evaluates the potential health risks to the human body resulting from contact with physical, chemical, or biological hazard factors. Exposure assessment involves determining the exposure pathways of pollutants, assessment models, model parameters, and population exposure analysis, among other aspects. The exposure dose depends on the concentration of pollutants in water, as well as the pathways, frequency, and duration of human exposure to pollutants (e.g., exposure parameters). Exposure pathways mainly refer to the ways of human intake, including inhalation, dermal absorption, and oral ingestion (Kaur et al. 2020).

The non-carcinogenic risk via the drinking water pathway (Liang et al. 2025), HQ_o can be assessed by first determining the daily exposure dose using the following equation (Eq. 9.2):

$$Intake_o = \frac{C \times IR \times EF_o \times ED}{BW \times AT} \tag{9.2}$$

Where, $Intake_o$ represents the daily exposure dose during long-term consumption of groundwater, mg/(kg·day). C denotes the concentration of the evaluated indicator, mg/L. IR is the daily water intake, L/d. EF_o stands for the exposure frequency, d/a. ED refers to the duration of water consumption, representing the number of years (a) during which the human body ingests a certain toxic substance. BW is the average body weight of humans, kg. AT is the average exposure time, a.

The daily exposure dose via dermal contact can be estimated using Eq. (9.3):

$$Intake_d = \frac{C \times AF \times ET \times CF \times SA \times EF_d \times ED}{BW \times AT} \tag{9.3}$$

Where, $Intake_d$ represents the exposure dose via dermal absorption, with the unit of mg/(kg·day). AF denotes the skin permeability coefficient, cm/h. ET is the exposure duration, h/d. EF_d stands for the exposure frequency through the dermal absorption pathway, d/a. SA refers to the skin exposure surface area, cm^2. CF is the conversion factor, with a value of 0.001.

The fourth step is risk evaluation. The health risk caused by each non-carcinogenic substance is expressed by HQ. Among them, the hazard quotient (HQ_o) resulting from human intake of the substance through the drinking water pathway is expressed as Eq. (9.4):

$$HQ_o = \frac{Intake_o}{RfD} \tag{9.4}$$

Where, RfD represents the reference dose for non-carcinogenic factors, in mg/(kg·day). Based on literature review, the reference dose for the non-carcinogenic effect of nitrate is 1.6 mg/(kg·day).

The entropy of harm through dermal contact, HQ_d is shown in Eq. (9.5):

$$HQ_d = \frac{Intake_d}{RfD} \tag{9.5}$$

The risk index *HI* to the human body caused by a single non-carcinogenic substance is the sum of the non-carcinogenic risks through different routes of intake and can be expressed by Eq. (9.6):

$$HI = HQ_o + HQ_d \tag{9.6}$$

The non-carcinogenic risk (*HI*) threshold value is 1.0. When $HI > 1.0$, it means that the degree of non-carcinogenic risk is high and the harm caused to human health is unacceptable, on the contrary, the non-carcinogenic health risk is acceptable.

9.4.1.2 Uncertainty Analysis

Exposure assessment serves as the quantitative basis for risk evaluation and constitutes a crucial link connecting dose-response assessment and risk characterization. In the above health risk assessment process, a specific value was adopted for the calculation parameters of exposure assessment. In fact, the selection of parameter values needs to be combined with specific events and the actual circumstances of the exposed population, leading to certain uncertainties. To further verify the rationality of the assessment results, Monte Carlo simulation was employed to evaluate the uncertainties in the health risk assessment (Huang et al. 2021). The Monte Carlo method primarily involves capturing the geometric quantities and characteristics of movement of objects, utilizing random sampling and statistical testing methods. Based on measured data, it constructs the probability process of events, samples from the probability distribution of random variables, and uses the generated random numbers as variable sequences for simulation to obtain approximate solutions to problems. The number of Monte Carlo simulations should be appropriate: an excessive number of simulations will increase computational time, while an insufficient number will result in a significant discrepancy between the random sequence and actual conditions, leading to large errors.

Monte Carlo simulation can be summarized into three steps:

1. Establishing probability distributions: Selecting the main factors that generate uncertainties and statistically analyzing the probability distributions of these influencing factors.
2. Distribution sampling: Using the factors with established probability distributions as variables, conducting simulation experiments to implement random sampling. The random numbers from random sampling serve as variable sequences and become their sampled values.
3. Establishing statistical quantity estimation: Collating the results of simulation experiments to obtain simulated values under different confidence levels, which are used as the simulation results of uncertainty analysis to analyze the problem.

9.4.1.3 Sensitivity Analysis

Sensitivity refers to the quantity of uncertainty in a forecast cell caused by the uncertainty of a hypothetical cell and model sensitivity. Figure 9.11 presents a diagram illustrating the relationship between sensitivity and uncertainty, which explains the impact of uncertainty and model sensitivity on two different predicted values under two scenarios. In each chart, the distance between the hypothetical variable and the leverage point represents the sensitivity of the model. A larger distance indicates a lower sensitivity of the model.

In the above model, although the assumed variables are assumed to have high uncertainty, they are not sensitive to the predicted values due to the low sensitivity of the model. In the following model, despite the low uncertainty of some assumed variables, they are highly sensitive to the predicted values because of the high sensitivity of the model. Calculating these complex relationships would be extremely difficult and time-consuming without programs like Crystal Ball.

The sensitivity of assumed variables affecting predicted values involves two factors: one is the model sensitivity of the assumed variables with respect to the predicted values, and the other is the uncertainty of the assumed variables. The sensitivity chart shows the degree of influence of each assumed variable cell on a specific predicted value. During the simulation, Crystal Ball ranks the assumed variables according to their importance to each predicted cell. The sensitivity chart presents this ranking in the form of bar graphs, indicating which assumed variables are the most or least important in the model. The Crystal Ball model calculates sensitivity by ranking the correlation coefficients between each assumed variable and each predicted value during the simulation runs. The correlation coefficient effectively measures the extent to which the assumed variable and the predicted variable change

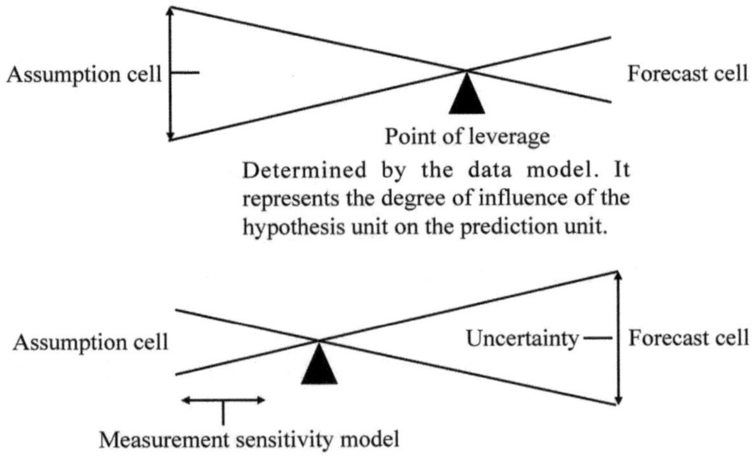

Fig. 9.11 Relationship between sensitivity and assumption uncertainty

together. If an assumed variable and a predicted value have a high correlation coefficient, it means that the assumed variable has a significant effect on the predicted value (including uncertainty and model sensitivity).

When calculating sensitivity, there may be inaccuracies in the calculation of relevant assumptions. If an important assumed variable is highly correlated with an unimportant one, the unimportant assumed variable may appear to have high sensitivity to the current predicted value. On the sensitivity chart, assumed variables are labeled according to their correlation. In some cases, removing correlations can help obtain more accurate sensitivity information.

A positive correlation coefficient indicates that as the assumed variable increases, the predicted value also increases. A negative correlation coefficient indicates the opposite. The larger the absolute value of the correlation coefficient, the stronger the correlation. The right and left sides of the 0 line in the sensitivity chart represent positive and negative correlations, respectively, and the length of the horizontal bars indicates the magnitude of sensitivity.

This study discusses the sensitivity of various parameters in the health risk assessment of nitrate for adults and children. The calculation of sensitivity generally defines the objective function as follows (Eq. 9.7):

$$K = F\left(x_1, x_2, \ldots, x_i, x_n\right) \tag{9.7}$$

Where, x_i represents the i-th parameter; n denotes the number of parameters. When all parameters change by Δx_1, Δx_2, …, Δx_i, respectively, K changes accordingly by ΔK, which indicates the total change in K caused by the combined variations of all parameters. It can be expressed using the Taylor expansion of the multivariate function (Eq. 9.8):

$$\Delta K = Kx_1 \Delta x_1 + Kx_2 \Delta x_2 + \ldots + Kx_n \Delta x_n \tag{9.8}$$

Where, Kx_i is the partial derivative of K with respect to x_i. If only the factor x_i changes (e.g., $\Delta x_i \neq 0$, $\Delta x_j = 0, j \neq i$), then ΔK_i, representing the impact of Δx_i on K, can be expressed as $\Delta K_i \approx Kx_{ii} \times \Delta x$. It is evident that $\Delta K = \Delta K_1 + \Delta K_2 + \ldots + \Delta K_i = \sum \Delta K_i$.

The sensitivity P_i of K to x_i is defined as the radio of the change in K to change in the factor x_i (Eq. 9.9):

$$P_i = \left(\Delta K_i / K\right) / \left(\Delta x_i / x_i\right) \tag{9.9}$$

Equation (9.9) implies that a 1% change in the i-th parameter x_i will lead to a P_i percent change in the objective function. A positive P_i indicates that ΔK_i and Δx_i change in the same direction. The larger the absolute value of $|P_i|$, the more sensitive the objective function K is to the parameter x_i. x_i can be regarded as a sensitive factor that plays a significant role in determining the value of the objective function K.

9.4.2 Deterministic Risk Assessment

By applying the exposure dose parameters and risk characterization parameters to the selected study area to Eqs. (9.2), (9.3), (9.4), (9.5) and (9.6), the health risk values associated with groundwater ingestion via both oral intake and dermal contact were derived for adults and children in the region. Detailed results are presented in Table 9.5.

In terms of exposure pathways (Table 9.5), the health risk indices of groundwater nitrate from oral ingestion for adults ranged from 9.72×10^{-4} to 6.03, with a mean value of 1.37. For children, the non-carcinogenic health indices via oral ingestion range from 2.27×10^{-3} to 14.06, with a mean value of 3.19. Regarding the oral ingestion pathway, 46.88% and 75.00% of the water samples exhibit hazard indices greater than 1 for the non-carcinogenic risks of groundwater nitrate for adults and children, respectively. This indicates that most groundwater is contaminated by nitrate, posing significant non-carcinogenic risks to human health. For the dermal contact pathway, the non-carcinogenic hazard indices are much less than 1, suggesting that the non-carcinogenic health risks caused by dermal contact are extremely low. The non-carcinogenic risks via oral ingestion are hundreds of times higher than those via dermal contact. The non-carcinogenic risk values under the combined exposure pathways are nearly identical to those under the oral ingestion pathway, indicating that oral ingestion is the primary exposure pathway contributing to non-carcinogenic risks to human health. Considering the combined results of oral ingestion and dermal contact, children face greater non-carcinogenic risks than adults. Based on the calculated non-carcinogenic health risk indices of groundwater nitrate for adults and children in the study area, the distribution of health risk indices was plotted (Fig. 9.12).

As can be seen from Fig. 9.12a, in the non-carcinogenic risk areas for adults, there are scattered water sampling points in the Mei County where the health index does not exceed 1, while the nitrate health risk indices throughout the Mei County are generally high. The health risk indices are low in the southwest and northeast of the Zhouzhi County, whereas they are high in other areas of the county. The health risk indices in the Huyi District are lower compared to those in the Mei County and Zhouzhi County. As shown in Fig. 9.12b, the distribution range of non-carcinogenic risk areas for children is wider than that for adults, covering the entire study area. Moreover, the risk indices in Mei County are the highest. The health risk indices in the Zhouzhi County are relatively lower than those in the Mei County, but the non-carcinogenic hazard indices in most areas of the Zhouzhi County are also higher than 1. The health risk indices in the northeast of the Huyi District are less than 1, indicating a low health risk.

Table 9.5 Deterministic health risk statistics for nitrate

Exposure pathways	Adults			Children		
	Max	Min	Mean	Max	Min	Mean
Oral intake	6.03	9.72×10^{-4}	1.37	14.06	2.27×10^{-3}	3.19
Dermal contact	8.87×10^{-3}	1.43×10^{-6}	2.01×10^{-3}	2.59×10^{-2}	4.17×10^{-6}	5.87×10^{-3}
Total	6.04	9.73×10^{-4}	1.37	14.09	2.27×10^{-3}	3.19

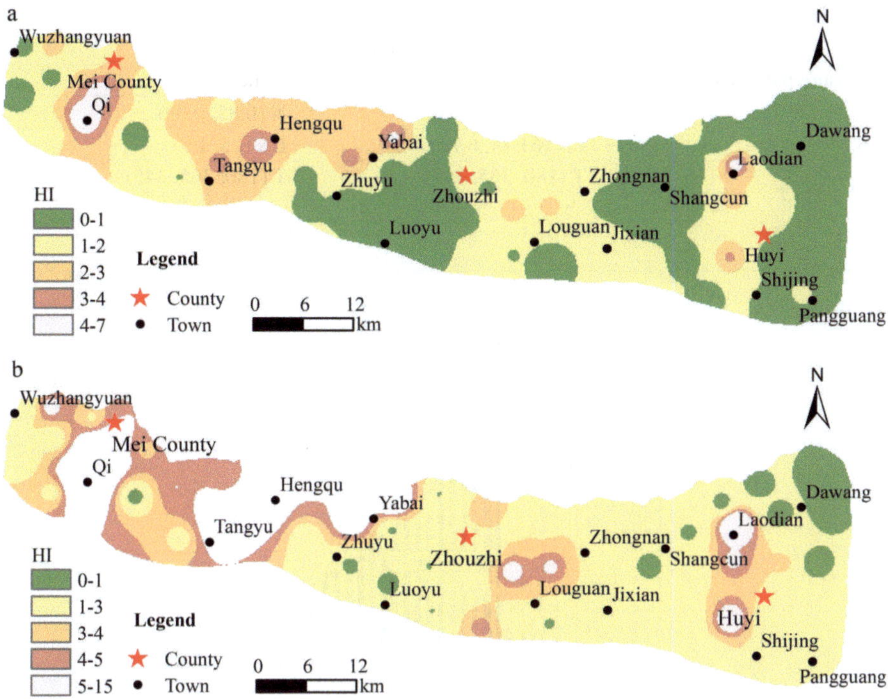

Fig. 9.12 Distribution of non-carcinogenic health risk indices for groundwater nitrate (**a**) adults; (**b**) children

9.4.3 *Stochastic Risk Assessment*

In this section, Crystal Ball risk simulation software was used to perform 10,000 iterations of stochastic simulation on the total health risk for adults and children. The risk levels are calculated and compared with the results of deterministic analysis. Based on the actual test values, the distribution type of nitrate concentration is determined to be lognormal distribution. The stochastic model parameters mainly consider indices such as intake rate (*IR*), body weight (*BW*), and exposure frequency (*EF*). Among them, *IR* follows a normal distribution, *BW* generally follows a log-normal distribution, and *EF* follows a triangular distribution. The distribution of nitrate concentration and the values of exposure parameters are shown in Table 9.6.

Figure 9.13 illustrates the stochastic simulation results of non-carcinogenic risks associated with groundwater nitrate. Red bars represent sampling points where the total non-carcinogenic risk value is less than the evaluation criterion of 1. Blue bars represent sampling points exceeding the acceptable non-carcinogenic risk threshold, indicating potential non-carcinogenic impacts on human health. The results of the stochastic model simulation show that the total non-carcinogenic risks in the study area all follow a log-normal distribution. In terms of non-carcinogenic risk probabilities, the exceedance probabilities of total non-carcinogenic risks for adults and children in the stochastic model are 42.56% (Fig. 9.13a) and 70.24% (Fig. 9.13b), respectively, which are not significantly different from the corresponding

Table 9.6 Range of values of exposure parameters

Exposure parameter	Distribution type (parameter)	
	Adults	Children
IR	N (1.5, 0.15)	N (0.85, 0.09)
BW	LN (61.75, 6.18)	LN (15, 1.5)
EF	T (180, 365, 345)	T (180, 365, 345)

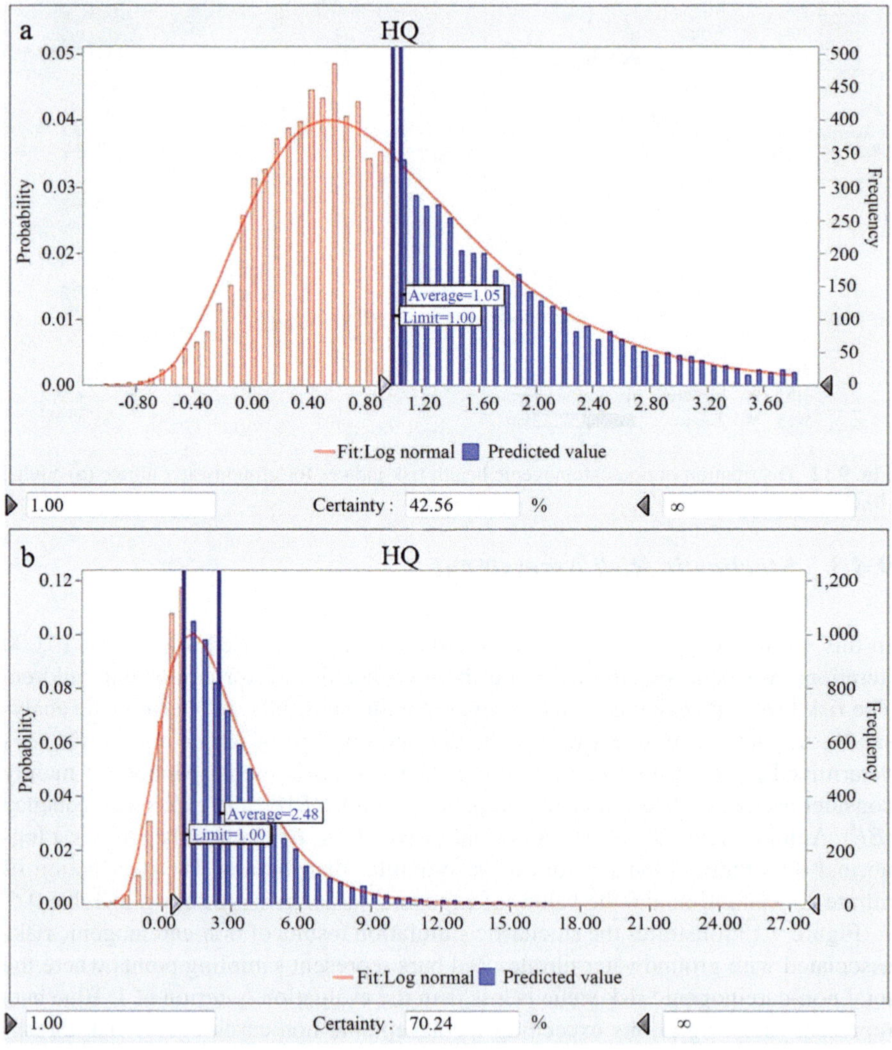

Fig. 9.13 Stochastic modelling results for groundwater nitrate non-carcinogenic risk. (**a**) adults; (**b**) children

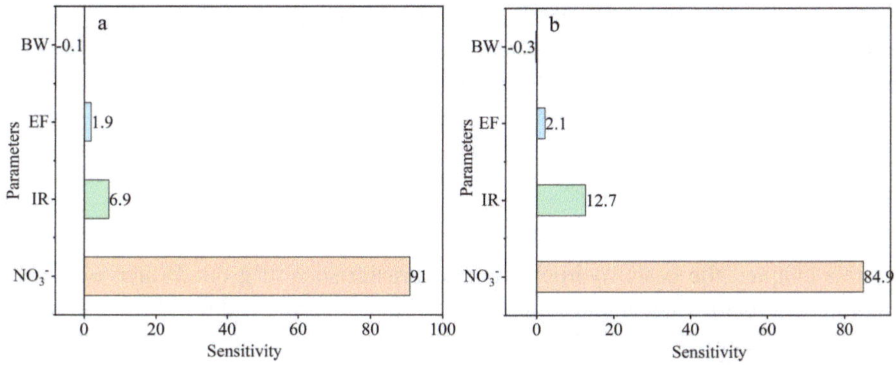

Fig. 9.14 Sensitivity analysis of non-carcinogenic risk of nitrate in groundwater: (**a**) adults; (**b**) children

exceedance probabilities in the deterministic model (46.88% and 75.00%, respectively). In terms of average values, the mean total non-carcinogenic risks for adults and children in the Monte Carlo stochastic model are 1.05 and 2.48, respectively. The corresponding mean values in the deterministic model are 1.37 and 3.19, showing an overall slight difference.

In addition, to determine the impact of multiple uncertain factors on the evaluation results, sensitivity analysis was performed on the health risk assessment results of the Monte Carlo stochastic model using Crystal Ball software, with the results presented in Fig. 9.14. A larger absolute value of sensitivity from the sensitivity analysis indicates a greater impact of the parameter on the risk assessment; positive and negative values represent a positive and negative correlation with the assessment results, respectively. Figure 9.14a, b depict the parameter sensitivities in the non-carcinogenic risks of nitrate for adults and children, respectively, revealing that non-carcinogenic risks are mainly influenced by nitrate concentration. Among the exposure assessment parameters, body weight is negatively correlated with health risks with low sensitivity, while the sensitivity of daily water intake (*IR*) is relatively higher than that of other parameters. Therefore, to ensure human health in the study area, priority should be given to mitigating groundwater pollution and controlling human intake of groundwater with excessive nitrate levels.

In summary, a comparison between the results of deterministic and stochastic risk assessments reveals that the mean values of the groundwater health risk assessment indices in the study area based on the Monte Carlo model are not significantly different from those derived from the deterministic model. Both models indicate that the total potential health risk faced by children is greater than that by adults. Additionally, most of the stochastic prediction indices are higher than the deterministic assessment indices, while the exceedance probability of stochastic health risk assessment is lower than that of the deterministic assessment. Both models can be utilized when appropriate assessment parameters are selected. However, stochastic assessment can present prediction results under different probability distributions, reflect uncertainties in risk assessment under varying conditions, and thus provide scientific decision-making basis for reducing stochastic risks in different situations.

9.5 Conclusions

In this chapter, the basic hydrochemical characteristics of groundwater were analyzed using mathematical and statistics and Piper diagram, and the current situation of nitrate pollution in the study area was clarified by examining the spatial distribution of nitrate. Based on the land use types derived from remote sensing interpretation, the differences in groundwater nitrate distribution under different land use patterns were analyzed. Isotope technology was employed to identify the sources of nitrate nitrogen in groundwater, and the MixSIAR model was further used for quantitative analysis of these pollution sources. Finally, deterministic health risk assessment and Monte Carlo-based stochastic health risk assessment methods were applied to analyze the non-carcinogenic health risks of groundwater nitrate in the study area. The specific conclusions are as follows:

1. The hydrochemical types of groundwater in the Qishan County and Mei County are both of HCO_3-Ca·Mg type. In Zhouzhi County and Huyi District, the groundwater is predominantly of the HCO_3-Ca·Mg type, with a small amount of SO_4·Cl-Ca·Mg type. The dominant ions in the groundwater of the study area are Ca^{2+} and HCO_3^-.
2. High concentrations of groundwater nitrate are mainly distributed in the Qi Town of Mei County, Yabai Town of Zhouzhi County, and Laodian Town of Huyi District, where the corresponding land use types are dominated by Farmland and Rural residential land. In contrast, in the southern part of the study area close to the Qinling Mountains, where the land type is mainly Grassland, the groundwater nitrate concentration is relatively low. Statistical analysis of groundwater nitrate concentrations across different land use types shows the following order: Farmland > Forest ≈ Urban and Rural residential land > Grassland > Water > Wasteland. Based on the results of land use types simulated by the FLUS model, it can be inferred that the groundwater nitrate concentration may continue to rise by 2040.
3. The results of the random forest model indicate that NO_3^- in groundwater is mainly influenced by TN, TDS, K^+, Ca^{2+}, and EC. The MixSIAR model was used to quantitatively calculate the contribution ratios of nitrate sources, revealing that soil nitrogen contributes the most to groundwater nitrate in farmland and Urban and Rural Residential areas within the region, accounting for 84.1% and 76.3% respectively. This is followed by chemical fertilizers, manure, and sewage.
4. The results of the deterministic health risk assessment show that the mean values of potential non-carcinogenic health risks faced by adults and children are 1.37 and 3.19, respectively. Moreover, the non-carcinogenic risk indices of both

groups exhibit a similar spatial distribution, with the highest values observed in the Mei County. Oral ingestion is the primary pathway causing human health risks in the study area, while the non-carcinogenic risks from dermal contact are extremely low.

References

Dash C et al (2024) Spatial variation of groundwater nitrate concentration in Guntur District, Andhra Pradesh, India: human health risk and mitigation strategy. In: Adhikary PP, Shit PK, Laha J (eds) Soil, water pollution and mitigation strategies. Environmental science and engineering. Springer, Cham. https://doi.org/10.1007/978-3-031-63296-9_16

Fida M, Li P, Wang Y, Alam SMK, Nsabimana A (2023) Water contamination and human health risks in Pakistan: a review. Expo Health 15(3):619–639. https://doi.org/10.1007/s12403-022-00512-1

Gao J, Wang S, Li Z et al (2021) High nitrate accumulation in the vadose zone after land-use change from croplands to orchards. Environ Sci Technol 55(9):5782–5790. https://doi.org/10.1021/acs.est.0c06730

Guo W, Li P, Du Q, Zhou Y, Xu D, Zhang Z (2024) Hydrogeochemical processes regulating the groundwater geochemistry and human health risk of groundwater in the rural areas of the Wei River Basin, China. Expo Health 16(2):291–306. https://doi.org/10.1007/s12403-023-00555-y

He S, Li P, Wu J, Elumalai V, Adimalla N (2020) Groundwater quality under land use/land cover changes: a temporal study from 2005 to 2015 in Xi'an, northwest China. Hum Ecol Risk Assess 26(10):2771–2797. https://doi.org/10.1080/10807039.2019.1684186

He S, Li P, Su F, Wang D, Ren X (2022) Identification and apportionment of shallow groundwater nitrate pollution in Weining Plain northwest China using hydrochemical indices nitrate stable isotopes and the new Bayesian stable isotope mixing model (MixSIAR). Environ Pollut 298:118852. https://doi.org/10.1016/j.envpol.2022.118852

Huang J, Wu Y, Sun J (2021) Health risk assessment of heavy metal(loid)s in park soils of the largest megacity in China by using Monte Carlo simulation coupled with positive matrix factorization model. J Hazard Mater 415:125629. https://doi.org/10.1016/j.jhazmat.2021.125629

Jahangir M, Johnston P, Barrett M et al (2013) Denitrification and indirect N_2O emissions in groundwater: hydrologic and biogeochemical influences. J Contam Hydrol 152:70–81. https://doi.org/10.1016/j.jconhyd.2013.06.007

Kaown D, Hyu Y, Bae G et al (2007) Factors affecting the spatial pattern of nitrate contamination in shallow groundwater. J Environ Qual 36(5):1479–1487. https://doi.org/10.2134/jeq2006.0361

Kaur L, Rishi MS, Siddiqui AU (2020) Deterministic and probabilistic health risk assessment techniques to evaluate non-carcinogenic human health risk (NHHR) due to fluoride and nitrate in groundwater of Panipat, Haryana, India. Environ Pollut 259:113711. https://doi.org/10.1016/j.envpol.2019.113711

Li P, He X, Li Y, Xiang G (2019) Occurrence and health implication of fluoride in groundwater of loess aquifer in the Chinese Loess Plateau: a case study of Tongchuan, northwest China. Expo Health 11(2):95–107. https://doi.org/10.1007/s12403-018-0278-x

Li S, Jiang H, Guo W et al (2023) From soil to river: revealing the mechanisms underlying the high riverine nitrate levels in a forest dominated catchment. Water Res 241:120155. https://doi.org/10.1016/j.watres.2023.120155

Li L, Li P, Tian Y, Kou X, He S (2024) Nitrate sources and transformation in surface water and groundwater in Huazhou District, Shaanxi, China: integrated research using hydrochemistry, isotopes and MixSIAR model. Environ Res 263:120052. https://doi.org/10.1016/j.envres.2024.120052

Li LX, Li PY, He S, Wang D, Tian Y, Niu L (2025) Groundwater nitrate pollution source apportionment under varying land use/land cover patterns. Expo Health 17:353–369. https://doi.org/10.1007/s12403-024-00666-0

Liang H, Li P, Elumalai V, Tian Y, Kou X (2025) Hydrochemical characteristics, groundwater nitrate sources and potential health risks in a typical alluvial plain of northwest China. Phys Chem Earth 139:103903. https://doi.org/10.1016/j.pce.2025.103903

Podgorski J, Labhasetwar P, Saha D et al (2018) Prediction modeling and mapping of groundwater fluoride contamination throughout India. Environ Sci Technol 52:9889–9898. https://doi.org/10.1021/acs.est.8b01679

Rivett M, Buss S, Morgan P et al (2008) Nitrate attenuation in groundwater: a review of biogeochemical controlling processes. Water Res 42(16):4215–4232. https://doi.org/10.1016/j.watres.2008.07.020

Rodriguez-Galiano V, Mendes M, Garcia-Soldado M et al (2014) Predictive modeling of groundwater nitrate pollution using Random Forest and multisource variables related to intrinsic and specific vulnerability: a case study in an agricultural setting (Southern Spain). Sci Total Environ 476–477(1):189–206. https://doi.org/10.1016/j.scitotenv.2014.01.001

Sattolo T, Mariano E, Boschiero B et al (2017) Soil carbon and nitrogen dynamics as affected by land use change and successive nitrogen fertilization of sugarcane. Agric Ecosyst Environ 247:63–74. https://doi.org/10.1016/j.agee.2017.06.005

Stock BC, Jackson AL, Ward EJ et al (2018) Analyzing mixing systems using a new generation of Bayesian tracer mixing models. PeerJ 6:e5096. https://doi.org/10.7717/peerj.5096

Tesoriero A, Gronberg J, Juckem P et al (2017) Predicting redox-sensitive contaminant concentrations in groundwater using random forest classification. Water Resour Res 53:7316–7331. https://doi.org/10.1002/2016WR020197

Vystavna Y, Diadin D, Grynenko V et al (2017) Determination of dominant sources of nitrate contamination in transboundary (Russian Federation/Ukraine) catchment with heterogeneous land use. Environ Monit Assess 189:509. https://doi.org/10.1007/s10661-017-6227-5

Wang S, Zheng W, Currell M et al (2017) Relationship between land-use and sources and fate of nitrate in groundwater in a typical recharge area of the North China Plain. Sci Total Environ 609:607–620. https://doi.org/10.1016/j.scitotenv.2017.07.176

Wang S, Zheng W, Kong X (2018) Spatial distribution characteristics of nitrate in shallow groundwater of the agricultural area of the North China Plain. Chin J Eco-Agric 26(10):1476–1482. https://doi.org/10.13930/j.cnki.cjea.180639

Wang D, Li P, Ji Y, Ren X (2025) Insight into the impact of biogeochemical reactions of groundwater nitrogen on chemical weathering and carbon cycling. J Hydrol 648:132458. https://doi.org/10.1016/j.jhydrol.2024.132458

Wei W, Nghiem A, Ma R et al (2021) Factors controlling iodine enrichment in a coastal plain aquifer in the North Jiangsu Yishusi Plain, China. J Contam Hydrol 243:103894. https://doi.org/10.1016/j.jconhyd.2021.103894

Xie Q, Chen Z, Khan A et al (2024) Organizational support, market access, and farmers' adoption of agricultural green production technology: evidence from the main kiwifruit production areas in Shaanxi Province. Environ Sci Pollut Res 31:12144–12160. https://doi.org/10.1007/s11356-024-31981-3

Xin J, Liu Y, Chen F et al (2019) The missing nitrogen pieces: a critical review on the distribution, transformation, and budget of nitrogen in the vadose zone-groundwater system. Water Res 165:114977. https://doi.org/10.1016/j.watres.2019.11

Chapter 10
Water Resources Protection and Management

Abstract This chapter presents a comprehensive assessment of current challenges and strategies for water resources protection and management in the northern Qinling Mountains. This ecologically vital region, while rich in surface and groundwater resources, faces mounting pressure from climate change, population growth, urbanization, and intensive land use. The chapter outlines key issues in water quantity, including declining runoff, groundwater overexploitation, and reduced ecological baseflow, and water quality with rising concerns over industrial, domestic, and agricultural pollution, particularly nitrate contamination. The ecological environment is further threatened by soil erosion, land degradation, habitat fragmentation, and recurrent natural hazards such as landslides and floods. In response, the chapter reviews legal frameworks, ecological policies, major infrastructure projects (e.g., Han-to-Wei and Hongyan–Shitou River transfers), and ecological restoration programs. It highlights integrated management approaches to coordinate surface water and groundwater use, restore damaged ecosystems, and mitigate pollution. The chapter concludes with forward-looking recommendations for sustainable water governance, emphasizing legal enforcement, scientific research, public participation, and coordinated conservation to ensure the Qinling Mountains' long-term ecological security and its role as China's central water tower.

Keywords Water resource management · Nitrate pollution · Ecological restoration · Climate change impact · Environmental policy

10.1 Existing Problems

The northern Qinling Mountains give rise to numerous rivers, forming a critical water source zone for the Guanzhong Basin. Numerous water conservancy projects have been constructed in the mountainous upper reaches of these rivers, and the intermittent storage and release of reservoir water have had considerable impacts on downstream aquatic ecosystems. In recent decades, rapid economic growth, population expansion, and accelerated urbanization in the Guanzhong Basin, along with

P. Li et al., *Water Resources Research in Northern Qinling Mountains, China*, Springer Hydrogeology, https://doi.org/10.1007/978-3-032-09395-0_10

concentrated human activities including industrial operations, agricultural fertilization, and domestic wastewater discharge, have led to notable pollution of rivers and groundwater, degradation of aquatic ecosystem health, and significant constraints on regional socio-economic development.

10.1.1 Water Quantity

In recent years, river runoff in the northern Qinling Mountains has exhibited a downward trend, as evidenced by declining annual runoff depths in three representative basins: the Hei River, Lao River, and Ba River (Jia et al. 2024). The primary driver of water volume reduction in the Qinling region is the climate change, especially declining precipitation. However, anthropogenic factors and land surface alterations have also contributed to it. Historical deforestation, hillside reclamation, and mining, along with rapid urban expansion and the increase in non-agricultural land use, have led to the extensive loss of forests, grasslands, lakes, wetlands, and paddy fields, significantly diminishing surface water area and the region's water retention capacity. The Weinan region exhibited the highest incidence of river flow interruptions. Field investigations identified seven rivers, Liuye, Changjian, Baimajian, Shuiyu, Yuxian, Shidi, and Gouyu Rivers with some segments experiencing complete flow cessation. At present, river discharges fall short of sustaining the ecological baseflows of major tributaries to the Wei River, including the Shitou, Hei, Lao, Chan, and Ling Rivers, as well as the environmental water needs of key lakes such as Kunming and Meibei. Climate-induced aridity plays a pivotal role in determining the reliability of ecological baseflow delivery. At the annual scale (1956–2016), the Hei River experienced the most pronounced reduction in annual runoff, followed by the Wangyu River (a tributary of the Ba River), whereas the Hao River (a tributary of the Jue River) exhibited no clear trend (You 2020). At the seasonal scale, all three rivers exhibited similar runoff trends: spring showed a marked decline, summer showed either a slight decrease or no clear trend, and autumn exhibited minimal or no change. However, winter runoff trends varied markedly across the three rivers: Hei River displayed a significant decline, Hao River a slight upward trend, and Wangyu River showed no statistically significant change (Fig. 10.1).

Groundwater is an important water source in the northern Qinling Mountains. Prior to 2000, over five decades of groundwater extraction had led to a general decline in groundwater levels, with particularly pronounced drops in key urban centralized water supply areas. Taking the Huyi District in the northern Qinling Mountains as an example, groundwater levels exhibited a fluctuating downward trend from 1980 to 2019. Interannual variations were more pronounced in alluvial and diluvial plains, but relatively modest at the margins of alluvial–diluvial fans. Key external drivers of groundwater level dynamics in the study area include

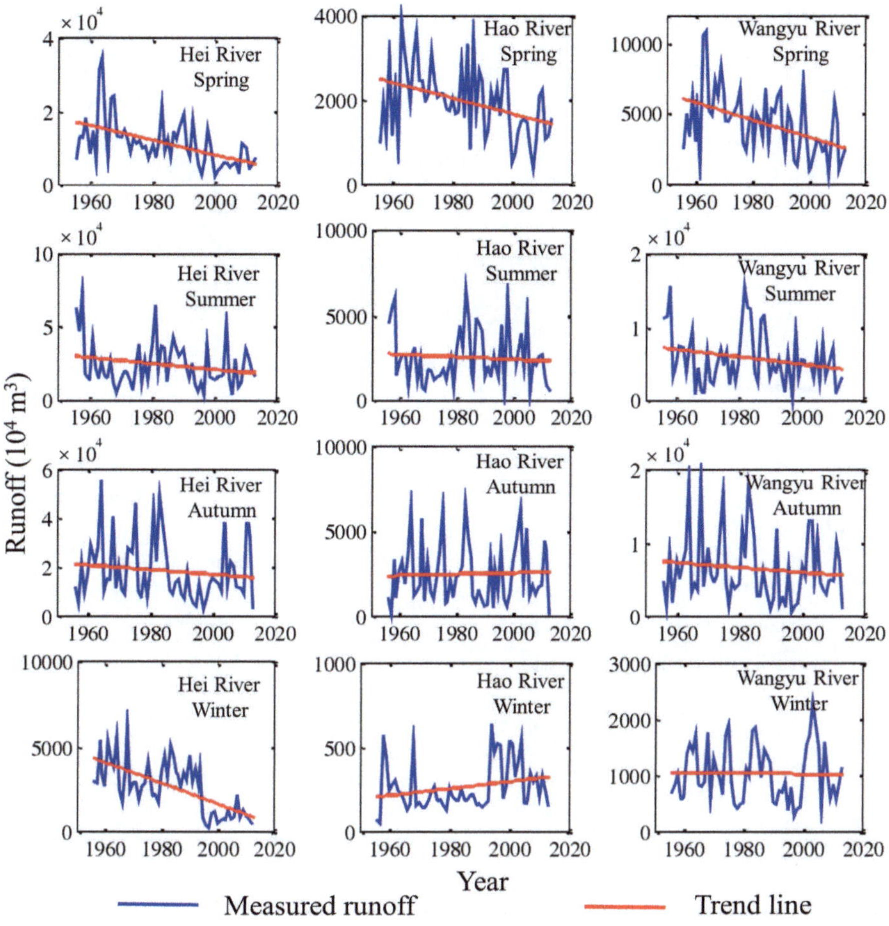

Fig. 10.1 Variation of runoff from Hei River, Hao River and Wangyu River. (Modified from You 2020)

precipitation, fluvial runoff, and groundwater extraction (Wang et al. 2020a). Climate change, coupled with prolonged overexploitation of groundwater, has triggered escalating environmental issues, including geohazards like land subsidence and surface fissuring.

Since 2000, the launch of the Hei River Water Conservancy Project enabled Xi'an to effectively overcome its longstanding water scarcity issues. With increased reliance on surface water, groundwater extraction declined, resulting in a moderated drop in groundwater levels across Xi'an. As showed in Fig. 10.2, the groundwater level in the northern Qinling Mountains has shown a significant upward trend from 2013 to 2023. In several urban zones, groundwater levels

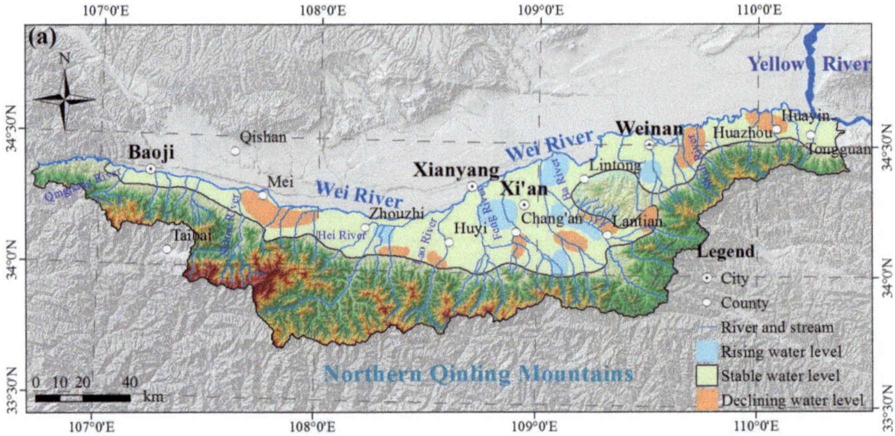

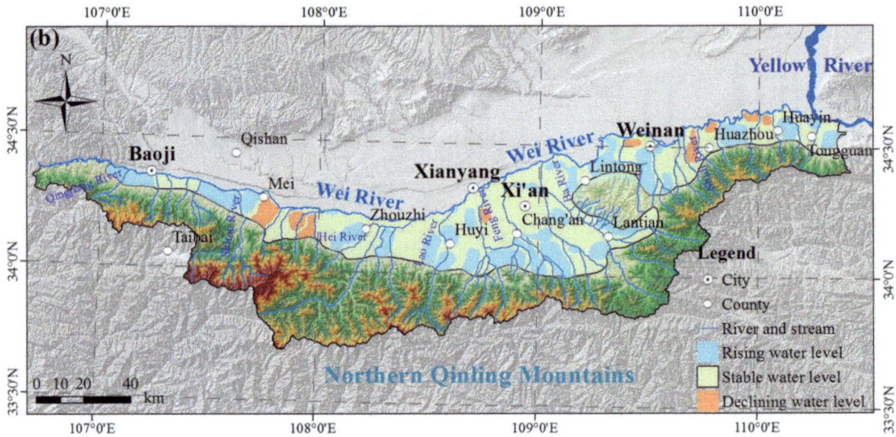

Fig. 10.2 Annual variation maps of groundwater depth: (**a**) 2013; (**b**) 2023

began to rebound, and confined aquifer levels exhibited notable recovery (Fig. 10.3). Due to heavier abstraction of groundwater in summer, the seasonal trend in phreatic groundwater levels typically shows a peak in the winter, a trough in the summer, with autumn levels slightly exceeding those in the spring. Confined groundwater levels follow a similar seasonal trend, peaking in the winter, reaching their lowest in the summer, with a marginal increase in the autumn compared to the spring. The annual average fluctuation range of confined groundwater is greater than that of phreatic groundwater (Yang et al. 2022). As groundwater levels continue to recover, integrated management of groundwater and surface water has emerged as a critical strategy for water resource governance in the northern Qinling Mountains.

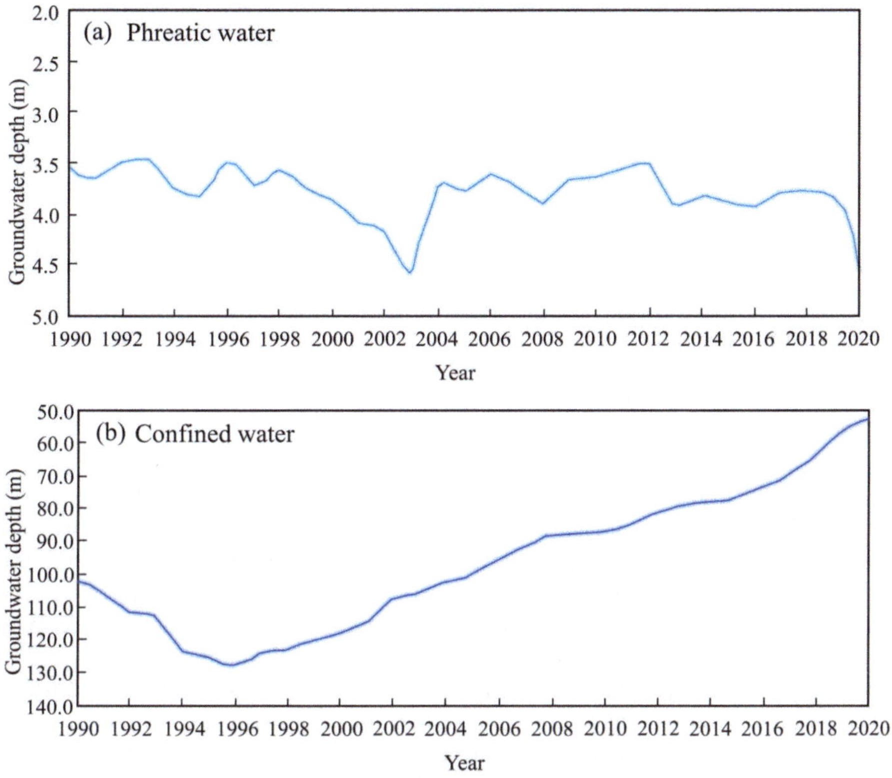

Fig. 10.3 Groundwater depth of typical monitoring wells in Xi'an City: (**a**) Phreatic water; (**b**) Confined water. (Modified from Yang et al. 2022)

10.1.2 Water Quality

River water quality in the northern Qinling Mountains exhibits a spatial gradient of degradation from upstream to downstream, with the upper reaches in Baoji City having better water quality than the lower reaches in Weinan City. Field investigations indicate that numerous factories remain in some valley areas, and the discharge of industrial wastewater, air pollutants, and solid residues poses a substantial threat to the water environment. Downstream of the valleys, water quality in Xi'an City and Weinan City is generally degraded, with ammonia nitrogen (NH_3-N) identified as the dominant pollutant. The conflict between economic development and environmental protection remains pronounced, leading to escalating pressure on the regional ecosystem. Taking the lower Wei River as an example, both NH_3-N and COD levels displayed a temporal trend of initial increase followed by a decline,

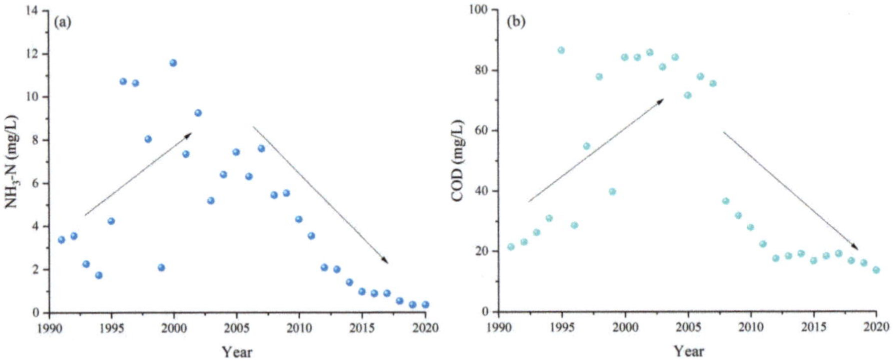

Fig. 10.4 Variation of NH_3-N (**a**) and COD (**b**) concentrations in the downstream of the Wei River. (Modified from Liu 2023)

with the inflection point generally occurring between 2002 and 2005 (Fig. 10.4). During this period, the Shaanxi provincial government initiated a comprehensive environmental remediation program for the Wei River in 2004. Notably, in 2005, Shaanxi launched the "Key Remediation Plan for the Wei River Basin" which emphasized systematic management, stringent pollution control, and source-level prevention to advance environmentally sustainable development (Liu 2023). Evidently, both NH_3-N and COD exhibited notable reductions in their multi-year average concentrations during the 2004–2020 remediation phase compared to the 1991–2003 pre-remediation period. These findings demonstrate a marked improvement in Wei River water quality following the implementation of remediation measures. The northern Qinling region serves as a major kiwifruit production base. Surface water in natural vegetation zones contains nitrate (NO_3-N) levels averaging 1.95–4.80 mg/L, whereas concentrations in intensive kiwifruit farming areas reach 16.12–36.28 mg/L, clearly indicating that kiwifruit cultivation contributes significantly to nitrate pollution in surface waters (Wang et al. 2020b).

Groundwater quality in the northern Qinling Mountains is generally favorable. Groundwater hydrochemistry is primarily controlled by the dissolution of silicate and carbonate rocks, cation exchange reactions, and agricultural inputs from human activities. Nitrate contamination and high TH are the principal factors contributing to substandard groundwater quality. Groundwater in the piedmont zone of the Qinling Mountains exhibits better quality compared to that in the surrounding plains, highlighting the urgent need for groundwater management and the reduction of agricultural non-point source pollution. The combined effects of agricultural runoff and domestic wastewater discharge in the plains have significant and often overlooked impacts on groundwater quality. Over time, ion concentrations in groundwater display site-specific trends across various monitoring wells. For instance, the Baoji monitoring well shows a slight upward trend in major ion concentrations. In contrast, wells in Xi'an and Weinan exhibit declining trends in ion concentrations (Fig. 10.5). The main kiwifruit cultivation zone in northern Qinling, faces significant groundwater nitrate contamination. Nitrate

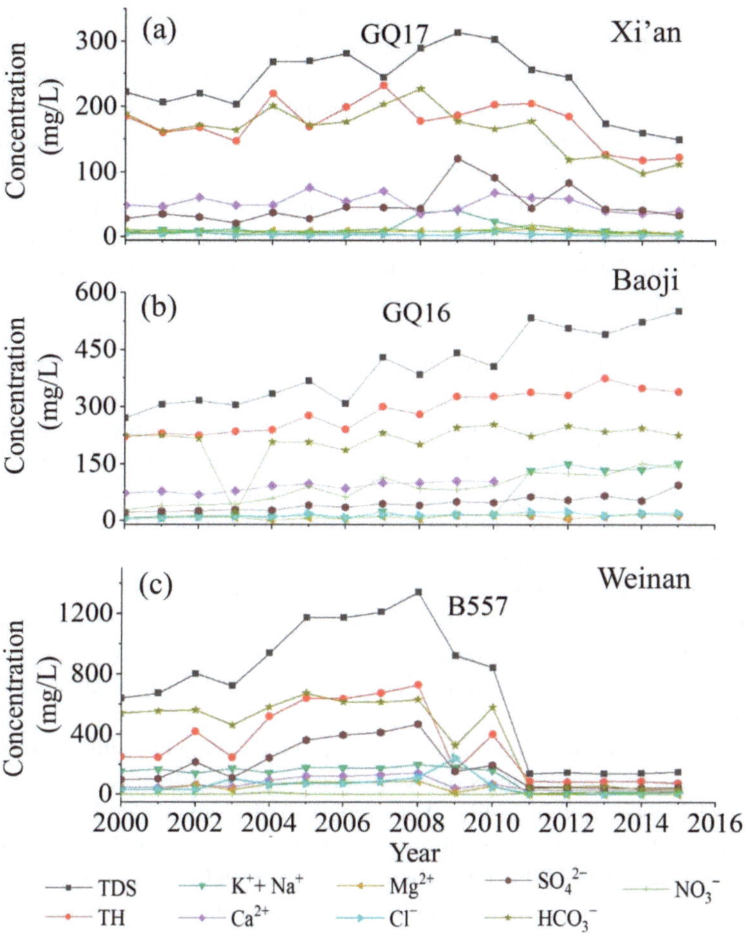

Fig. 10.5 Chemical concentration evolution of three monitoring wells in Xi'an City (**a**), Baoji City (**b**), and Weinan City (**c**). (Modified from Nsabimana 2023)

pollution is most severe in shallow aquifers, with mean concentrations reaching 29.68 mg/L (NO_3-N) and an exceedance rate as high as 93.27%. For deeper groundwater aquifers, the exceedance rate is 57.41%, with an average nitrate concentration of 15.06 mg/L (Wang et al. 2020b).

10.1.3 Ecological Environment

The northern Qinling Mountains are characterized by steep slopes, pronounced elevation gradients, active faulting, and complex geomorphology, which create ideal conditions for landslides and debris flows. These unique regional conditions result in frequent torrential rain, dramatic river stage fluctuations, and recurrent flash floods, contributing to the high frequency and broad distribution of geological disasters. These disasters lead to annual losses of life and property, significantly constraining the region's socioeconomic development. In recent years, climate variability has intensified, marked by frequent extreme weather events. Persistent heavy rainfall is expected to aggravate these risks further. Intensive resource extraction and vegetation loss have accelerated soil erosion, affecting nearly 50% of the Qinling area and resulting in an annual soil loss of approximately 84 million tons (Chen and Zhang 2019). Alterations in land use have disrupted the spatial distribution of soil organic matter, posing significant threats to the sustainability of agricultural practices. Despite ongoing ecological conservation efforts, the region's economy remains underdeveloped and reliant on land-intensive industries, highlighting the sharp tension between environmental protection and rural development. With limited arable land and high dependency on forest resources, illegal logging and wildlife poaching are persistent concerns. Unregulated and excessive tourism development poses a significant threat to the integrity of wildlife habitats. Human-induced degradation of forest resources persists in the northern Qinling region, presenting major challenges for ecological protection. In remote forested areas of northern Qinling, local communities often lack a fundamental understanding of the importance of ecological stewardship. Pursuit of short-term economic gains has led to forest destruction via illegal logging for mushroom and charcoal production, land clearance, and steep-slope farming. These practices have triggered a cascade of environmental problems.

The role of water conservation in ecosystems is receiving growing attention from both ecologists and hydrologists due to its ecological significance. Water retention represents a key ecosystem service that supports hydrological stability and ecological resilience. The water conservation capacity in the northern Qinling Mountains declined during 1980–2000, but rebounded during 2000–2020. Spatially, water retention is higher in the southern highlands and lower in the northern alluvial plains (He et al. 2025). From 1985 to 2020, water retention in the northern Qinling area rose at an annual rate of 1.10 mm, with climate change explaining 86.3% of the increase and land cover change accounting for the remaining 13.7% (Wang et al. 2024). Thus, it remains to be further explored whether the water conservation function of vegetation in the Qinling Mountains has approached its ecological threshold.

10.2 Protection and Management Strategies

10.2.1 Legal Policies and Management Systems

As awareness of the need to protect the Qinling Mountains has increased, the Shaanxi provincial government and its agencies have progressively introduced a range of legislative and regulatory measures. Key policies include the *Shaanxi Provincial Regulations on the Ecological Protection of the Qinling Mountains*, the *Master Plan for the Qinling Ecological Conservation*, and the *Implementation*

Table 10.1 Policy documents concerning the protection of the Qinling Mountains

Policy documents	Release date	Release department
Resolution of the CPC Shaanxi Provincial Committee on Enhancing the Ecological Protection of the Qinling Mountains	2018	CPC Shaanxi Provincial Committee, People's Government of Shaanxi Province
Shaanxi Provincial Regulations on the Ecological Protection of the Qinling Mountains	2019	Standing Committee of the Shaanxi Provincial People's Congress
Qinling Mountains Ecological Protection Action Plan	2019	People's Government of Shaanxi Province
Measures for Fully Leveraging Procuratorial Functions to Strengthen Ecological and Environmental Protection of the Qinling Mountains	2019	People's Procuratorate of Shaanxi Province
Master Plan for the Qinling Ecological Conservation	2020	General Office of the Shaanxi Provincial Government
Reward Mechanism for Reporting Ecological and Environmental Destruction in the Qinling Mountains	2021	Shaanxi Provincial Development and Reform Commission
Special Plan for the Protection and Utilization of Water Resources in the Qinling Mountains	2021	People's Government of Shaanxi Province
Special Plan for Pollution Prevention and Control in the Qinling Mountains	2021	Shaanxi Provincial Department of Ecology and Environment
Industry Access List for Core and General Protected Areas in the Qinling Mountains	2023	Shaanxi Provincial Development and Reform Commission
Implementation Guidelines for Strengthening Qinling Ecological Protection and Promoting High-Quality Development	2024	General Office of the CPC Shaanxi Provincial Committee, General Office of the Shaanxi Provincial Government
Wetland Protection Plan of Shaanxi Province (2025–2030)	2025	Shaanxi Provincial Forestry Bureau

Guidelines for Strengthening Qinling Ecological Protection and Promoting High-Quality Development, among others. Further details are provided in Table 10.1. These policy documents serve as a crucial foundation for the effective conservation and management of the Qinling Mountains' ecological environment. Nevertheless, the effective execution of numerous policy documents has become a new and pressing challenge.

10.2.2 Project Measures

In recent years, inter-basin water diversion projects have been actively planned and implemented to mitigate water scarcity on the northern Qinling Mountains. Notable examples include the Han River-to-Wei River Diversion Project and the Hongyan River–Shitou River Water Transfer Project. The Han River-to-Wei River Project is a strategic and large-scale water infrastructure initiative linking the Han and Wei River basins, designed to support integrated development across Guanzhong, southern, and northern Shaanxi, while safeguarding the province's ecological and socio-economic stability. The project will meet the domestic and industrial water needs of 21 recipient regions, including four major cities (Xi'an, Xianyang, Weinan, Yangling), 11 counties/districts along the Wei River, five towns in the Xi-Xian New Area, and the Weibei Industrial Park (Gaoling, Lintong, and Yanliang clusters), while returning the water originally diverted from agriculture. The project will significantly alleviate groundwater over extraction and ecological water encroachment in the Guanzhong region, helping to restore groundwater balance and reduce the risk of urban geo-environmental hazards. It will serve the daily and industrial water needs of approximately 14.11 million people living along the Wei River corridor.

The Hongyan–Shitou River Water Transfer Project, situated in Taibai County, Baoji, extracts water from the upper Hongyan River, a Bao River tributary on the southern Qinling Mountains, and diverts it via a long gravity tunnel through the mountain range into the Shitou River, a tributary of the Wei River on the northern slope. Following regulation by the Shitou River Reservoir, the diverted water will supply major cities including Xi'an, Xianyang, Baoji, and Yangling, while also supplementing ecological flows to the main stem of the Wei River. The system is engineered to divert water at a peak rate of 13.5 m^3/s, transferring up to 92.4 million m^3 annually. Post-regulation by the Shitou River Reservoir, the project will provide up to 266 million m^3 of water annually, with 46.96 million m^3 designated for ecological supplementation to the Wei River. The project is critical in optimizing spatial allocation of water across Shaanxi, enhancing water security for the Guanzhong urban cluster, and mitigating conflicts among urban-rural supply, development demand, and ecological needs.

10.2.3 Restoration and Governance

Since the early 2000s, unauthorized villa construction in the northern Qinling Mountains has involved deforestation and terrain alteration, elevating landslide hazards. Concurrent wastewater discharge has contaminated water sources, severely threatening the local ecosystem. In response, Shaanxi Province launched a targeted campaign to identify and address unauthorized villa construction and other ecological disturbances on the northern Qinling Mountains. A total of 1194 illegal villas were identified: 1185 were legally demolished, and 9 were confiscated. The government reclaimed 3 km^2 of state-owned land and returned 2 km^2 of collective land, achieving comprehensive demolition and confiscation. After demolishing the illegal villas, vegetation was fully restored and the ecosystem gradually improved.

Recently, Shaanxi Province launched an Integrated Protection and Restoration Project of Mountains, Rivers, Forests, Farmlands, Lakes, Grasslands, and Deserts in the Northern Qinling Mountains, Shaanxi. This project is structured into seven pillars: groundwater recharge, biodiversity conservation, water yield enhancement, land and geological rehabilitation, erosion control, watershed-level restoration, and smart management systems. To date, Xi'an has achieved 350 km^2 of ecological restoration and 668.85 km of river rehabilitation, which significantly enhanced the ecological quality of the Xi'an segment of the northern Qinling Mountains.

Long-term industrial and agricultural activities have led to nitrate contamination, posing a significant threat to groundwater safety in the northern Qinling Mountains. Denitrification is the primary mechanism responsible for reducing NO_3^- concentrations in groundwater (Wang et al. 2023). However, in the plain areas of the northern Qinling region, denitrification is relatively limited; for instance, the denitrification potential in the major kiwifruit-producing zone of Zhouzhi–Mei County ranges from 6.99 to 15.92%, which is insufficient to markedly lower NO_3^- concentrations in soils and groundwater (Wang et al. 2023; Liu et al. 2025). As a result, nitrate leaching and accumulation persist, further worsening groundwater contamination. The limited attenuation capacity, compounded by sustained NO_3^- inputs due to excessive fertilization in agricultural zones, intensifies the severity of this issue. Effective groundwater nitrate mitigation requires two-pronged measures: controlling nitrogen input (e.g. through optimized fertilizer and manure management) and reducing contamination risks in hydrogeologically vulnerable areas (Liu et al. 2025). Targeted remediation should be implemented in areas with severe nitrate contamination in groundwater.

10.3 Future Prospects

In recent years, heightened awareness of the Qinling Mountains' ecological significance has led to notable advancements in water resource protection and management on their northern slope, resulting in improved water quality and progressive ecosystem recovery. In the future, continued efforts in water resource protection and governance are essential to advancing ecological civilization and ensuring the sustainable use of water resources and long-term ecological improvement in the northern Qinling region (Li et al. 2025). Preserving the Qinling ecosystem is a long-term and challenging endeavor that demands coordinated action from government agencies, industry, academia, and the public. By refining legal frameworks, advancing scientific innovation, and fostering public engagement, the current challenges can be effectively tackled, paving the way for the sustainable development of the Qinling ecosystem. Looking ahead, the principles of science-based protection and integrated governance should be upheld, with continuous innovation to improve the effectiveness of conservation strategies. Strengthening scientific research on Qinling's aquatic ecosystems and establishing baseline data on ecological and health-related geology in the northern foothills are vital.

Water supply and demand structures should be optimized to enhance resource allocation efficiency. Agricultural, industrial, and urban water management should be reinforced to raise efficiency across sectors and improve overall resource utilization. Enhanced water quality surveillance, including refined monitoring protocols, and strict control over non-point source pollution and emissions from agriculture and industry are essential. Innovative approaches to ecosystem restoration and governance should be actively pursued. Public education and participation must be enhanced to cultivate a stronger awareness of water environment protection. Improving early warning systems and predictive capabilities for droughts and floods will support ecological disaster risk reduction and mitigation.

A symbiotic and co-evolving system integrating mountains, water, forests, farmland, lakes, and grasslands should be established in the northern Qinling region. Scientific research should inform policy-making to reinforce the role of the Qinling Mountains as China's "central water tower."

References

Chen Y, Zhang H (2019) Ecological environment of the Qinling Mountains: past, present and future. Chinese Science News 2019-03-07(008). (In Chinese)

He S, Qian H, Liu Y et al (2025) Water conservation assessment and its influencing factors identification using the InVEST and random forest model in the northern piedmont of the Qinling Mountains. J Hydrol Reg Stud 57:102194. https://doi.org/10.1016/j.ejrh.2025.102194

Jia L, Yu K, Li Z et al (2024) Spatiotemporal variation of potential evapotranspiration in the Qinling Mountain Areas and attribution analysis of typical watershed runoff. Trans Chin Soc Agric Eng 40(24):59–71. (in Chinese with English Abstract)

Li X, Li P, Liu J (2025) Water security in the Qinling Mountains: current status, climate and human influences, and adaptive strategies. Int J Water Resour Dev. https://doi.org/10.1080/07900627.2025.2478387

Liu G (2023) Characteristics of water quality changes in the Shaanxi section of the Weihe River main stream and evaluation of the effect of its control policy implementation. Master thesis, Xi'an University of Architecture and Technology. (in Chinese with English Abstract)

Liu Y, Qian H, Gao Y et al (2025) Identifying nitrate pollution of groundwater based on aquifer vulnerability and surface nitrogen loads in the Northern Foot of the Qinling Mountains, China. J Hazard Mater 495:138939. https://doi.org/10.1016/j.jhazmat.2025.138939

Nsabimana A (2023) Groundwater quality in the Guanzhong Basin, China: novel approaches of comprehensive appraisals, prediction, and management. PhD thesis, Chang'an University

Wang H, Duan L, Wang W (2020a) Dynamic features of groundwater level in northern Qinling and its influence factors. Northwest Geol 53(2):280–288. (in Chinese with English Abstract)

Wang S, Qu T, Hu H et al (2020b) Dynamic water quality changes in a kiwifruit production area of the northern Qinling Mountains. J Agro-Environ Sci 39(12):2853–2859. (in Chinese with English Abstract)

Wang D, Li P, Yang N et al (2023) Distribution, sources and main controlling factors of nitrate in a typical intensive agricultural region, northwestern China: vertical profile perspectives. Environ Res 237(1):116911. https://doi.org/10.1016/j.envres.2023.116911

Wang F, Wang W, Wu Y, Li W (2024) Assessment of water retention dynamic in water resource formation area: a case study of the northern slope of the Qinling Mountains in China. J Hydrol Reg Stud 56:102063. https://doi.org/10.1016/j.ejrh.2024.102063

Yang C, Tao F, Yuan X (2022) Analysis of groundwater dynamic characteristics in major basin areas of Shaanxi Province. Northwest Geol 55(3):345–354. (in Chinese with English Abstract)

You Q (2020) Study on the hydrologic variation of the main rivers at the northern foothill of Qinling Mountains supplied water to Xi'an City and the safety of urban water supply. PhD thesis, Chang'an University

Chapter 11
Conclusions and Future Perspectives

Abstract This chapter summarized the main findings from this book, and proposed some potential research perspectives and possible research topics that may foster the protection and management of water resources in the North Qinling Mountains. This concluding chapter synthesizes the key findings from the book, which have systematically investigated the hydrology, water quality, ecological sustainability, and human impacts on the water resources of the northern Qinling Mountains. The synthesis highlights the critical role of this region as China's "central water tower" and the pressing challenges it faces from climate change and anthropogenic pressure. Building upon these conclusions, this chapter outlines a forward-looking research agenda aimed at addressing persistent knowledge gaps. This agenda is explicitly aligned with global priorities, including the United Nations' Sustainable Development Goals (SDGs), and emphasizes the transformative potential of emerging technologies such as artificial intelligence (AI) and the inclusive framework of citizen science. The perspectives offered here are intended to guide future interdisciplinary research and evidence-based policy for the sustainable management of water resources in this vital region.

Keywords Water resource management · Artificial intelligence · Citizen science · Climate change impact · Environmental policy

P. Li et al., *Water Resources Research in Northern Qinling Mountains, China*, Springer Hydrogeology, https://doi.org/10.1007/978-3-032-09395-0_11

11.1 Conclusions

This book has provided a comprehensive and interdisciplinary examination of the water resources in the northern Qinling Mountains, a region of paramount importance for water security in the Guanzhong Plain and beyond. The findings from this book reveal a complex interplay between natural hydrological processes and intensifying human activities, offering both insights and warnings.

1. **Hydrological System and River Dynamics**: Our research cataloged and analyzed the extensive river network originating from the northern Qinling slopes. We identified two distinct river types based on geomorphology and recharge mechanisms, which dictate their vulnerability to change. The ecological baseflow assessment revealed that many small- and medium-sized rivers, particularly in the eastern (Weinan) section, experience seasonal drying or possess measured flows below the calculated ecological requirements. This is driven by a combination of spatially uneven precipitation, unfavorable local hydrogeology, and upstream water interception by reservoirs and human use. The health of these rivers is the first indicator of the basin's overall water stress.
2. **Water Quality Status and Challenges**: The assessment of surface water and groundwater quality presents a mixed picture. While many upstream and reservoir waters are of good quality (Class I–III), significant degradation was observed downstream areas. Key issues include nitrate pollution from intensive agriculture and industrial activities, salinization, and microbial contamination. The spatial analysis clearly links deteriorating water quality to land-use patterns, with pollution hotspots correlated with urban expansion, agricultural intensification, and specific industrial points. This degradation threatens both ecosystem health and human water security.
3. **Critical Interactions and Human Impacts**: The investigation into surface water–groundwater interactions quantified recharge dynamics and revealed that river seepage is a significant recharge source in some areas, but this also acts as a pathway for pollutants entering aquifers. Land use change analysis demonstrated a clear trend of urbanization and agricultural expansion at the expense of natural vegetation, directly impacting the hydrological cycle by reducing infiltration and increasing pollution runoff. A stark case study on nitrate pollution in groundwater explicitly links its prevalence to specific land-use changes and quantifies the associated non-carcinogenic health risks, particularly for children. This chain of evidence directly connects human activities on the land to the quality of water in the rivers and aquifers and, ultimately, to public health.
4. **Towards Integrated Management**: Synthesizing these findings, some integrated water resource management (IWRM) strategies were proposed. These strategies are not merely technical but are essential policy recommendations focused on protecting ecological baseflows, controlling pollution at its source, implementing managed aquifer recharge (MAR), and adopting adaptive water allocation policies that prioritize ecological health alongside human needs.

In summary, the northern Qinling Mountains, while still a vital water source, is under severe stress. Its water resources are finite, vulnerable to quality degradation, and unevenly distributed. Sustainable development in the Guanzhong Plain is inextricably linked to the health of this mountain ecosystem. The findings of this book underscore the urgent need for a paradigm shift from exploitative to protective management, grounded in robust scientific evidence.

11.2 Future Perspectives

To build upon the foundation laid by this book and address the evolving challenges, future research must be interdisciplinary, innovative, and globally relevant. The following perspectives outline some promising directions.

11.2.1 Research Aligned with Sustainable Development Goals (SDGs)

Future research in the Qinling Mountains holds great potential to advance multiple United Nations Sustainable Development Goals (SDGs), thereby enhancing the region's global significance and informing policies with broader international impact.

- **SDG 6 (Clean Water and Sanitation)**: Research should focus on developing cost-effective, nature-based solutions (NBS) for wastewater treatment and pollution mitigation in rural and peri-urban areas. Studies on the economic valuation of water ecosystem services can strengthen the argument for conservation investments.
- **SDG 13 (Climate Action)**: A critical gap is understanding the resilience of the "central water tower" to climate change. Future work must develop high-resolution climate-hydrological models to predict changes in precipitation patterns, snowmelt, extreme droughts, and floods, and assess their impact on water availability.
- **SDG 15 (Life on Land)**: Research is needed to quantify the ecological thresholds of key species to water stress and pollution. This involves defining the "safe operating space" for water extraction to prevent irreversible damage to biodiversity, thus strengthening the ecological dimension of integrated water resources management.
- **SDG 2 (Zero Hunger) and SDG 12 (Responsible Consumption and Production)**: The clear link between agriculture and water quality calls for research into sustainable agricultural practices. This includes optimizing water-saving irrigation technologies, promoting precision agriculture to reduce fertilizer use, and evaluating the water footprint of different cropping systems in the region.

11.2.2 Harnessing Technological Innovation: AI and Big Data

Artificial intelligence (AI) has become a transformative force in science and society, reshaping lifestyles and creating new opportunities for sustainable development. In environmental research, AI offers unprecedented capabilities for analyzing vast, complex datasets, detecting hidden patterns, and making accurate predictions (He et al. 2025; Alotaibi and Nassif 2024). Applied to the Qinling Mountains, AI and machine learning (ML) can revolutionize water resources research and management.

- **AI-Driven Predictive Modeling**: ML algorithms can integrate multi-source data (hydrological, meteorological, geological, and land-use information) to develop robust models that forecast water quality changes, groundwater fluctuations, and flood risks with higher accuracy than conventional methods. This allows for proactive decision-making in water resource planning and disaster prevention. However, predictive models often require long-term, high-quality datasets, which may be sparse or inconsistent in the Qinling region. Additionally, AI models can act as "black boxes," making their results difficult to interpret for policy and management decisions. Therefore, future work should be done to establish standardized, long-term monitoring networks to improve data quality, promote explainable AI approaches to enhance transparency, and encourage interdisciplinary collaborations to ensure model outputs are usable by hydrologists, ecologists, and policymakers.
- **Pollution Source Apportionment and Early Warning**: AI can analyze complex water quality datasets to identify, quantify, and trace pollution sources with greater accuracy than traditional statistical methods. Coupled with real-time sensor networks, AI can deliver early warnings for pollution incidents, such as nutrient surges or algal blooms in reservoirs, enabling timely mitigation. However, sensor networks may face technical and financial barriers in rural or mountainous regions. Furthermore, AI-based apportionment models can be sensitive to incomplete datasets, leading to uncertainty in pollution source identification. As such, investment in low-cost, durable water quality sensors adapted to mountainous environments, creation of open-access pollution databases and integration of AI outputs into early-warning platforms that are linked directly to local management agencies should be encouraged.
- **Data Fusion from Remote Sensing**: AI excels at processing the vast amounts of data generated by satellites, drones, and wireless sensor networks. This enables high-frequency monitoring of land-use change, vegetation health (a proxy for water stress), and even fine-scale detection of point-source pollution. Such integration provides a holistic understanding of watershed dynamics. Whereas, remote sensing data can be limited by cloud cover, terrain complexity, or resolution constraints, and AI-based data fusion methods may amplify existing biases if not carefully validated. Moreover, access to high-resolution satellite imagery may be restricted due to cost or licensing issues. Promote partnerships for open and affordable access to high-resolution imagery, develop hybrid approaches that combine AI with ground-based validation, and implement transparent protocols

to reduce bias and enhance reliability in AI-driven remote sensing applications are necessary to address these challenges.

AI has the potential to transform water resources research in the Qinling Mountains by enabling predictive modeling, pollution source identification, and advanced remote sensing analysis. Yet, these benefits come with challenges related to data availability, model interpretability, cost, and accessibility. Overcoming these obstacles requires coordinated actions such as strengthening monitoring networks, investing in digital and sensing infrastructure, ensuring open data access, and fostering cross-disciplinary collaboration. Only through such measures can AI be fully harnessed to support sustainable water management and conservation in this ecologically vital region.

11.2.3 Embracing Citizen Science for Inclusive Monitoring

Citizen science presents a powerful tool to expand monitoring reach and foster public engagement (Dickinson and Bonney 2012). Citizen scientists can help gather environmental observations across vast areas and over extensive periods, which is often too costly or logistically complex for professional teams alone. This scale of data collection empowers deeper insights and long-term analysis. Qinling Mountains is huge, and thus citizen science may be helpful in fostering the research in this area.

- **Expanding Spatial and Temporal Coverage**: Engaging local communities, schools, and non-governmental organizations (NGOs) in simple water quality monitoring (e.g., using portable turbidity tubes, nitrate test strips) or reporting on river flow status can generate valuable data at a scale impossible for researchers alone. This is particularly useful for monitoring the numerous small rivers identified in this book.
- **Enhancing Public Awareness and Stewardship**: Involving citizens in data collection fosters a deeper connection to local water resources and a better understanding of environmental challenges. This can build crucial public support for conservation policies and sustainable water use practices.
- **Crowdsourcing Environmental Data**: Developing mobile applications for citizens to report observations, such as illegal dumping, pipe discharges, or river drying, can create an interactive, real-time map of environmental issues, providing valuable ground-truthing for official monitoring.

11.2.4 Interdisciplinary and Systems-Based Approaches

Addressing the complex and interconnected environmental challenges in the Qinling Mountains requires transcending traditional disciplinary boundaries and adopting holistic, systems-based approaches. Research must embrace methods such as water–energy–food nexus studies (Rai 2023) and social–hydrological systems modeling (Viola et al. 2021).

- **Water–Energy–Food Nexus Studies**: water–energy–food nexus studies integrate water, energy, food, and ecosystem considerations to reveal synergies and trade-offs that single-sector analyses often overlook. Future research should adopt a nexus approach to understand the trade-offs and synergies between water use for agriculture, urban supply, hydropower, and ecosystem needs. This is essential for formulating policies that avoid solving one problem by exacerbating another.
- **Social-Hydrological Systems Modeling**: Social-hydrological modeling embeds human behaviors, policy decisions, and socio-economic feedbacks directly into hydrological models, offering more realistic simulations and management strategies that reflect the dynamic interplay between society and water systems. Therefore, integrating social sciences into hydrological models is crucial. Research needs to explore the feedback loops between human decision-making (e.g., farmers' water use, policy responses) and hydrological processes to create more realistic and effective management scenarios.

The northern Qinling Mountains stand at a crossroads. The path forward requires a commitment to transformative science and collaborative governance. By leveraging new technologies, engaging the public, and aligning research with global sustainability goals, we can ensure that this "central water tower" continues to sustain life, economies, and ecosystems for generations to come.

References

Alotaibi E, Nassif N (2024) Artificial intelligence in environmental monitoring: in-depth analysis. Discov Artif Intell 4:84. https://doi.org/10.1007/s44163-024-00198-1

Dickinson JL, Bonney R (2012) Citizen science: public participation in environmental research. Cornell University Press, New York

He S, Zhou X, Liu Y, Zhao X, Guan Z, Ji Y, Li P (2025) Machine learning techniques in hydrogeological research. In: Li P et al (eds) Sustainable groundwater and environment: challenges and solutions. Springer, Cham, pp 137–164. https://doi.org/10.1007/978-3-031-82194-3_7

Rai PK (2023) Role of water–energy–food nexus in environmental management and climate action. Energy Nexus 11:100230. https://doi.org/10.1016/j.nexus.2023.100230

Viola F, Caracciolo D, Deiddaa R (2021) Modelling the mutual interactions between hydrology, society and water supply systems. Hydrol Sci J 66(8):1265–1274. https://doi.org/10.1080/02626667.2021.1909729

Index

P. Li et al., *Water Resources Research in Northern Qinling Mountains, China*, Springer Hydrogeology, https://doi.org/10.1007/978-3-032-09395-0

MIX
Papier aus verantwortungsvollen Quellen
Paper from responsible sources
FSC® C105338

If you have any concerns about our products,
you can contact us on
ProductSafety@springernature.com

In case Publisher is established outside the EU,
the EU authorized representative is:
Springer Nature Customer Service Center GmbH
Europaplatz 3, 69115 Heidelberg, Germany

Printed by Libri Plureos GmbH
in Hamburg, Germany